# SpringerBriefs in Geography

More information about this series at http://www.springer.com/series/10050

Qian Zhang · Xiangzheng Deng

# Urban Development in Asia: Pathways, Opportunities and Challenges

 Springer

Qian Zhang
Institute of Geographic Sciences
  and Natural Resources Research
Chinese Academy of Sciences (CAS)
Beijing
China

Xiangzheng Deng
Institute of Geographic Sciences
  and Natural Resources Research
Chinese Academy of Sciences (CAS)
Beijing
China

ISSN 2211-4165                          ISSN 2211-4173   (electronic)
SpringerBriefs in Geography
ISBN 978-981-10-2895-3                  ISBN 978-981-10-2896-0   (eBook)
DOI 10.1007/978-981-10-2896-0

Library of Congress Control Number: 2016954548

Printed on acid-free paper

This Springer imprint is published by Springer Nature
The registered company is Springer Nature Singapore Pte Ltd.
The registered company address is: 152 Beach Road, #22-06/08 Gateway East, Singapore 189721, Singapore

# Preface

China has experienced an unprecedented urbanization process over the last two decades, characterized by a high rate of urban land expansion and rural–urban migration. Recently, urbanization has been regarded as the most important economic engine for boosting China's development. Some predictions suggest that a hundred million people will move from the countryside to cities by 2020. Although China's government has launched an initiative of a new-type urbanization plan and an eco-civilization construction guideline, the most sustainable and smart way forward for urban development in China is still unclear. This includes where, when, and what types of cities should be built across the country in the next several years. Therefore, it is imperative to explore options for the most suitable urban development pathway for China.

Other countries' experiences of urbanization processes, challenges, and opportunities, and the strategies used to respond these situations, can provide valuable lessons for China. However, urbanization pathways in different countries are distinctly diverse on the account of their unique historical, political, economic, cultural, and religious backgrounds. Consequently, there is no simple way or single indicator to compare the urbanization pathways for different countries.

The purpose of this work is to provide an overview of urbanization pathways in selected Asian countries (namely, Japan, South Korea, Russia, India, Indonesia, Israel, and Pakistan), and to analyze the characteristics of their urban development. For these countries, we analyze and compare the temporal sequence of the urbanization process, urban spatial layout, urban scale structure, dynamic mechanisms, industrial sector employment, the level of economic development/activity, urban–rural relations, and the coordination of development. On the basis of the analysis and comparison, we generate implications for China's urban development. The experiences of other countries and the lessons gained therefrom are able to shed light on China's new-type urbanization.

The authors claim full responsibility for any errors appearing in this work.

Beijing, China
June 2016

Qian Zhang
Xiangzheng Deng

# Acknowledgments

This work was financially supported by the External Cooperation Key Program of the Chinese Academy of Sciences (Grant No. 131A11KYSB20130023) and the National Natural Science Funds of China (Grant No. 71533004, 71561137002). We also thank our project team for their knowledgeable contribution and unwavering support throughout the production of these results. Without their contribution this work would not have been published. In particular, we thank the participants of the summer school on sustainable urban development and international comparison, which was organized by the Faculty of Resources and Environmental Science, Hubei University in July, 2014. We acknowledge Ms. Dongdong Chen, a Master's student at Hubei University, for constructing all the graphs in this book. Finally, we are grateful to Springer staff for their enduring enthusiasm in helping us produce this work.

We are grateful to the authors of the numerous books and research publications mentioned in the list of references at the end of each chapter of this work. This valuable array of literature formed the foundation of the present work. We express our gratitude to those researchers and organizations for their contributions that reinforced our knowledge.

# Acknowledgments

# Contents

# List of Figures

# List of Tables

# Urban Development in Asia: Pathways, Opportunities and Challenges

## 1 Introduction

Urbanization has proved to be an inevitable feature of the development of humankind, and an indispensable part of national modernization, closely connected to economic development (Jedwad and Vollrath 2015). In different historical stages, countries have developed various modes of urbanization according to different political influences, economic and cultural conditions, and social systems (Jiang and O'Neill 2015). China, for example, is currently undergoing a process of accelerated urbanization. This process can be understood and informed by combining the universal laws of urbanization with the national conditions of the country, as well as by learning from the experience and lessons of urbanization in other countries. A comparative study of different countries' urbanization has significance for guiding the healthy and sustainable development of China's future urbanization (Chauvin et al. 2016). However, urbanization in China has encountered a series of economic and social problems, making it the focus of contradictions in the nation's modernization. Moreover, one of the keys to China's modernization lies in the timely implementation of urbanization strategy. Therefore, it is very important to devise appropriate strategies for guiding and controlling urbanization in China's socialist modernization and to implement them appropriately (Xin 2014).

Urbanization in the Asian region is a good reference point for China. China is located in the eastern part of Asia, and borders 14 other countries: North Korea, Russia, Mongolia, Kazakhstan, Kyrgyzstan, Tajikistan, Afghanistan, Pakistan, India, Nepal, Bhutan, Myanmar, Laos, and Vietnam (Fig.1.1). It is imperative to compare urbanization between different Asian countries because not only does the development of large cities and other urban areas play an important role in promoting the economic development of the region, but also the urbanization pathways and strategies have different effects in these countries (Seto et al. 2011). The development of cities continues to have a strong influence on the region's social

© The Author(s) 2017
Q. Zhang and X. Deng, *Urban Development in Asia: Pathways, Opportunities and Challenges*, SpringerBriefs in Geography, DOI 10.1007/978-981-10-2896-0_1

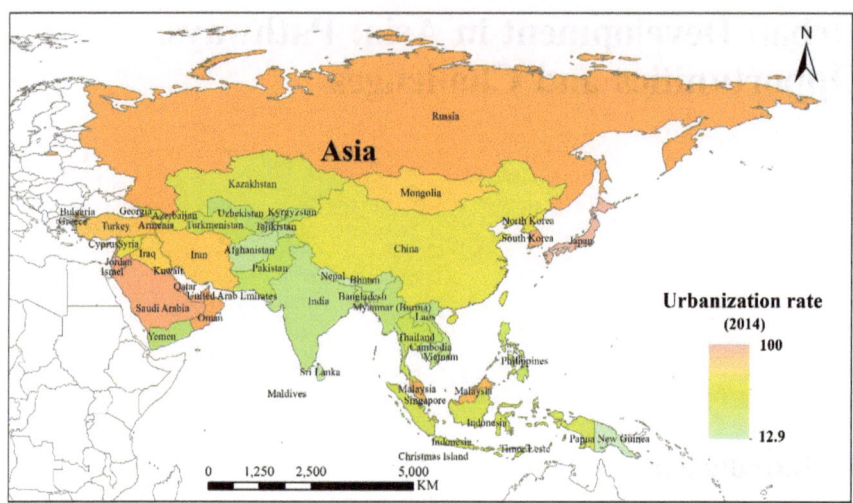

**Fig. 1.1** Urbanization rates of Asian countries (*Data source* World Bank)

and economic development, adjustments in industrial structure, and modernization (Henderson and Wang 2007).

According to the different levels of urbanization in Asia, we analyzed typical countries from North Asia, South Asia, West Asia, East Asia, and Southeast Asia. The countries selected for comparison are Japan, South Korea, Russia, India, Indonesia, Israel, and Pakistan. Japan, South Korea, and Israel are representatives of developed Asian countries with high population densities, and have experienced rapid urbanization, leading to urbanization rates (the percentage of people living in urban areas) in excess of 90 %. India is similar to China in that it is a country with a large and important emerging economy, has a huge population, and is undergoing very rapid urban development. Russia was selected as one of the comparison countries because of its unique urbanization process and social institutional system. Pakistan, as a less developed country, a critical node in the China–Pakistan Economic Corridor, and part of the Belt and Road Initiative, has great potential to be substantially urbanized in the next few decades. Indonesia was selected to represent the developing island nations of Asia. These seven perspectives are used to analyze and compare the characteristics of Asian urbanization: the temporal sequence of the urbanization process, urban spatial layout, urban scale structure, dynamic mechanisms, industrial sector employment, the level of economic development/activity, urban–rural relations, and the coordination of development.

# References

Chauvin JP, Glaeser E, Ma Y, Tobio K (2016) What is different about urbanization in rich and poor countries? Cities in Brazil, China, India and the United States. J Urb Econ

Henderson JV, Wang HG (2007) Urbanization and city growth: the role of institutions. Reg Sci Urb Econ 37:283–313

Jedwad R, Vollrath D (2015) Urbanization without growth in historical perspective. Explor Econ Hist 58:1–21

Jiang L, O'Neill BC (2015) Global urbanization projections for the shared socioeconomic pathways. Glob Environ Change

Seto KC, Fragkias M, Guneralp B, Reilly MK (2011) A meta-analysis of global urban land expansion. PLoS ONE 6(8):e23777

Xin Y (2014) International urbanization: comparative studies and implication. National School of Administration Press, Beijing (in Chinese)

# 2    Urban Development in Selected Asian Countries

## 2.1    Japan

### 2.1.1    Overview of Japan

Japan is an island country located in the eastern part of Asia and consists of four large islands (Honshu, Shikoku, Kyushu, and Hokkaido) and more than 6900 small islands. The population is 126 million and is dispersed over a land area of 377,900 km$^2$. In 2014, Japan's urbanization rate was 92.5 % (i.e., 92.5 % of the population lives in urban areas, where "urban area" is defined as a city of more than 50,000 persons and the percentage of industrial and commercial employees is more than 60 % of the total number of employees). Being a mountainous country, Japan is divided into two parts by a medial mountain range system, with one part bordering the Pacific Ocean and the other bordering the Japan Sea. Mountains and hills account for 71 % of the total land area, with forests covering around 67 % of the total area. Mount Fuji is Japan's highest peak, with an altitude of 3776 m. Japan's flat land is distributed mainly in the downstream parts of river systems and in coastal areas, mostly small-scale alluvial plains. The overall area of such plains is small, and therefore the area able to be devoted to agriculture is very limited. The population density is 2924 people/km$^2$, ranking the nation twenty-sixth most densely populated in the world. Japan has created the largest area of reclaimed land in the world.

In 2014, Japan's gross domestic product was US\$4.6 $\times$ 10$^{12}$, ranking it third in the world, and the per capita gross domestic product (GDP) was US\$36,000 (NBSC 2015). Japan's world-leading service industries, especially banking, finance, shipping, insurance, and business services, make the largest contribution to the GDP. The capital, Tokyo, is not only the largest city and economic center of the country, but also the largest city in Asia and the city with the highest gross value of production in the world, as well as being the world's top financial, shipping, and service center.

Japan's industry is characterized by having a full range of industrial types, advanced production technology, large scales of production, and high output value. The major industrial sectors include steel, automobiles, shipbuilding, electronics, chemicals, and textiles. Japan's agriculture is highly mechanized and efficient, with the yield per unit of cultivated land being the world's highest. The nation is self-sufficient in rice, but about half of the overall food for consumption needs to be imported. Japan is the world's second-largest fish-resource-producing country (Fig. 2.1).

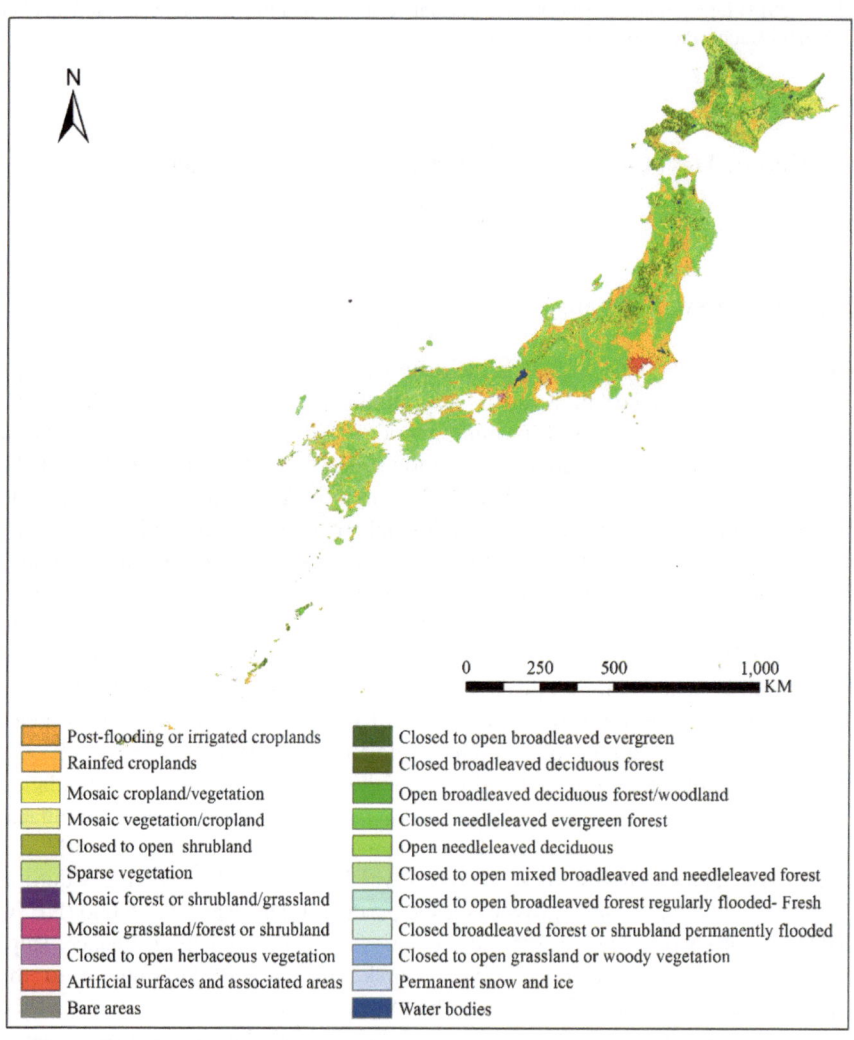

**Fig. 2.1** Land use map of Japan (*Data source* European Space Agency 2009)

## 2.1.2  The Urbanization Process of Japan

Japan's industrialization began in the Meiji Restoration and drove the process of urbanization. However, until 1940, the level of urbanization still lagged behind that of industrialized European and American countries. After World War II, the Japanese economy grew dramatically and the urbanization rate rose rapidly. In just 30 years, Japan went through an urbanization and industrialization process that had taken European and American countries more than 100 years to complete. Japan's urbanization can be divided into four main stages, as discussed below.

### The first stage: the preparatory phase of Japan's industrialization and urbanization (1868–1920)

Japan's industrialization and urbanization originated in the Meiji Restoration starting in 1868. In this stage, a series of policies were implemented to promote the development of agriculture. First, a series of bans that were enacted in the Tokugawa Shogunate period were abolished and the old feudal land ownership relations were reformed, creating a favorable environment for agricultural development and the transfer of rural labor into industrial production. Second, the "abolition of the system" (substituting county bureaucracies with Bakuhan institutions) and the establishment of the city, town, and village system changed previous territorial statuses and provided institutional support for the development of cities. Third, the government carried out a vigorous program of constructing industrial infrastructure, and began to build a nationwide railway network, as well as formulating preferential policies for the mining, steel, shipbuilding, and machinery industries. At the same time, light industry also developed rapidly, with the textile industry becoming a pillar industry. The collection of large amounts of agricultural tax formed the original capital for industrial development. In this stage, agriculture was still the dominant economic activity in Japan, accounting for 69.9 % of all employment in 1888, leaving 30.1 % for other industrial sectors. In 1920, the percentage of Japanese urban area accounted for only 0.4 % of the total land area and the urbanization rate was 18 %.

### The second stage: the transition stage of Japan's urbanization (1920–1945)

Industry developed rapidly in this stage, promoting a substantial transfer of the rural population into cities, in particular into heavy industrial cities. Four well-known industrial zones, namely, Keihin Industrial Zone, Chukyo Industrial Zone, Hanshin Industrial Zone, and North Kyushu Industrial Zone, formed industrial pillars. However, large numbers of workers remained in rural areas after World War II (WWII), and this slowed the process of urbanization, with the nation's urbanization rate remaining at 37.1 % until 1945.

## The third stage: accelerated development stage of Japan's urbanization (1945–1980)

This stage represented Japan's period of recovery after WWII. In this stage, Japan continued to strengthen economic development and urban restoration work, increasing the demand for labor, with population continuing to flow from the rural areas into the cities in response to this demand. The four major industrial areas formed before WWII and the three metropolitan areas centered on Tokyo, Kobe, and Nagoya provided favorable conditions for urbanization after WWII.

In this stage, the secondary industrial sector was predominant, and industrial zones and metropolitan areas developed related industries, promoting employment and attracting high numbers of workers into cities. In this stage, 429,000 workers transferred from the rural labor force to urban (chiefly manufacturing) employment. In addition, after World War II, the surplus that developed in the rural labor force to some extent accelerated the flow of rural population into cities. In 1965, primary industry employment represented only about 24.7 % of the workforce, secondary industry employment rose to 31.5 %, and tertiary industry employment rose to 43.8 %. The urbanization rate reached 67.9 %, with the urban area accounting for 23.5 % of the total land area in Japan. Urban growth in Japan followed a compact pattern as the urbanization rates in different stages are higher than the percentage of urban land as the total land area, which is not the general pattern for global cities. However, the rapid urbanization also brought a series of problems, including a large number of young people flowing into big cities, making city populations become too concentrated and regional disparities larger. It was not until 1975 that the conditions of Japanese inhabitants in Tokyo, Osaka, and Nagoya tended to improve with the slowdown of industrial growth and the maturing of the urbanization process.

## The fourth stage: maturity of industrialization and urbanization and re-urbanization stage (1980–present)

The period starting in the 1980s represents the peak development stage of Japan's urbanization. After the 1970s, Japan entered the post-industrial era and tertiary industry gradually became the leading industry of the national economy. In 1975, Japan's urbanization rate was 75.9 %, whereas in 2000 it had risen only slightly to 78.7 %, signifying that urbanization had essentially reached saturation. This mainly reflected the blurring of the boundaries between urban and rural areas, whereby the phenomenon of urban residents moving into rural areas with good-quality environments and cheap housing meant that rural and urban workforces became mixed together spatially, indicating the rise of small towns and the trend of "reverse urbanization" (deurbanization) at and beyond the fringes of large cities (Martinez-Fernandez et al. 2016). In addition, Japanese urban living standards became higher and the quality of urban infrastructure gradually improved, meaning that urban citizens became more satisfied with their living conditions over time (Figs. 2.2 and 2.3).

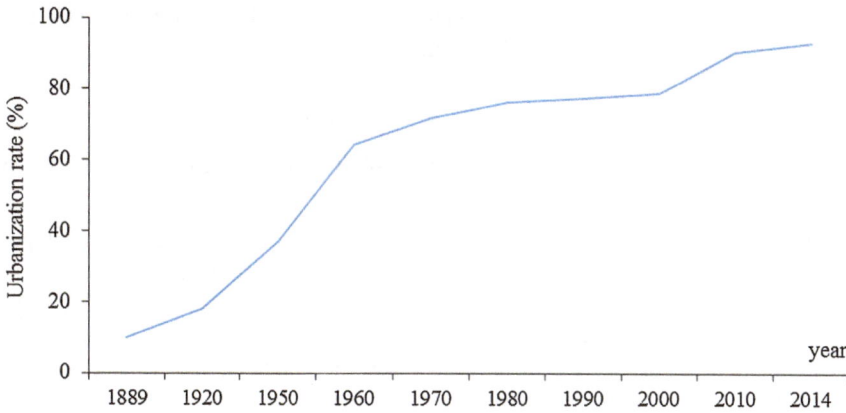

**Fig. 2.2**  Japan's urbanization trend (*Data source* World Bank)

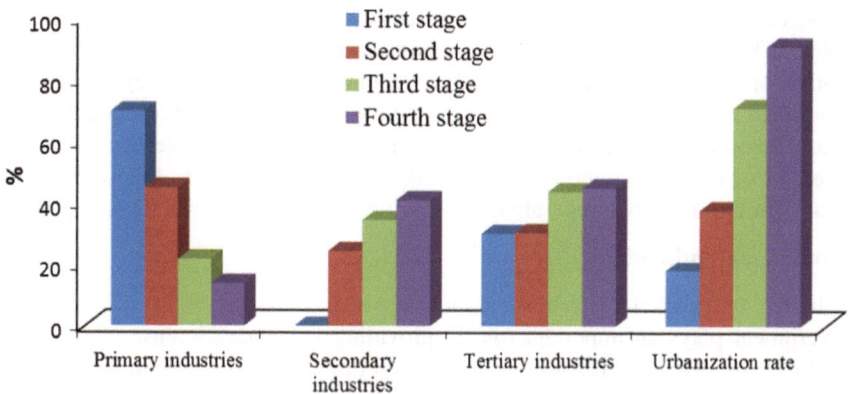

**Fig. 2.3**  Trends of the primary, secondary, and tertiary industrial sectors and of the urbanization rate for stages 1–4 of Japan's urbanization development (*Data source* World Bank)

### 2.1.3   Characteristics of Japan's Urbanization

#### The mutual promotion of industrialization and urbanization

Japan's urbanization developed from its start in the early twentieth century to approach maturity in the 1970s, and during this period industrialization developed alongside urbanization. From the 1950s to the 1970s, secondary industries contributed about 50 % of Japan's GDP, making those decades the golden age of industrialization, and the nation's urbanization developed rapidly in sympathy with industrialization from the 1950s to the 1970s. In 1955, Japan's urban population was 50,530,000, accounting for 56.1 % of the total population, after which time Japan's industrialization entered its accelerated growth phase. In the 1960s, the average growth rate of the Japanese economy reached 10.1 %, and that decade was also the period during which the development of urbanization in Japan accelerated. From 1963 to 1973, the average rural labor force working in non-agricultural sectors was 800,000 people, nearly 60 % of the rural labor force.

**The urbanization pattern became concentrated then dispersed, but on the whole is still relatively concentrated**

An important feature of Japan's urbanization process is that large cities developed first and then spread out to surrounding areas and even to other cities, a feature that is also connected with the background to Japan's industrialization. The Meiji government ushered in a capitalist society. However, at that time, Japan had neither a capitalist class nor any industrial base when the capitalist society started to be established, and therefore the government promoted the early industrialization of Japan by relying on feudal bureaucrats and the power of the chaebols. This created the situation whereby industrialization was initiated in large and medium-sized cities, where politicians and businessmen were concentrated, and then gradually radiated out to surrounding areas, eventually forming a metropolitan circle of Tokyo, Osaka, and Nagoya. However, from a nationwide perspective, the industrialization of Japan has shown a relatively centralized model. Japanese industry is distributed mainly in the area extending from the edge of Tokyo Bay along the Seto Inland Sea coast directly to Kitakyushu, with Hanshin in the middle. Japan has 656 cities, 11 of which have more than one million people, and most of these cities are located in the industrial zone along the Pacific coast, except for Sapporo.

**Government guidance**

Although the market mechanism dominates in the process of urbanization, the government plays an important role in providing guidance (Sorensen 2016). For example, the market works as a major force in the allocation of resources, with the government revising industrial policy when the market allocation fails. While the rural population was flowing into cities, the Japanese government paid attention to the establishment of integrated management systems, improved the rural labor transfer system to provide vocational skills training for the rural labor force, and modified laws and regulations to protect rural interests.

### 2.1.4  Challenges for Japan's Urban Development

**Industrial public nuisance, over-crowded housing, and traffic congestion**

Because of the speed of urbanization and industrial development in Japan, cities' populations became too dense, giving rise to some public nuisance problems and over-crowding. As early as the beginning of the Meiji period, several major mining pollution events occurred, such as those in Tongshan, the Hitachi mines (exhaust pollution), and the Besshi copper mine (exhaust pollution). Of the urban problems, the most difficult to solve was the housing problem caused by an over-concentrated population. After the Meiji period, a large number of rural inhabitants migrated to Tokyo, Osaka, and other major cities, resulting in over-crowding and forcing the poorest people to live in slums. Even at the peak of industrialization in the 1960s, a significant number of people were living in relatively poor residential conditions. Subsequently, although housing conditions improved, they could not be compared with those found in Europe or the United States, particularly with respect to the relatively small, cramped houses in Japanese cities. Cramped housing coupled with high prices forced a large number of people to move to the cheaper outskirts, which in turn exacerbated traffic congestion in the major cities (Lan and Zhang 2013).

**The price of urban land soared, threatening the economic development of Japan**

Land availability became a fundamental problem in Japanese cities. Problems associated with housing, transportation, and the natural and urban environments were all essentially caused by limited space. A shortage of land causes land prices to increase, a simple function of supply and demand. However, in Japan, this phenomenon became particularly exacerbated. In the 17 years from 1955 to 1972, Japanese land prices rose by 17.5 times, a far greater increase than occurred in other countries. This constituted one of the major influences on the creation of the subsequent "bubble economy," which in turn caused the long-term depression of the Japanese economy.

**The cities became overloaded and the residential living environment deteriorated**

The rapid development of the economy led to accelerated urbanization. At the same time, residents' incomes and consumption levels also rose sharply. However, accompanying this, residential living environments became worse. With the rapid development of industrialization and urbanization, the discharge of pollutants from industrial and mining-based activities using oil as the main fuel increased dramatically, as did automobile emissions, making the air quality worse. In large cities with high population densities, such as Tokyo, Osaka, and Yokohama, there were dozens or even hundreds of photochemical smog alerts every year. In addition, the amount of urban waste increased dramatically because of the over-crowding, and garbage-handling problems occurred, such as the event known as the "Tokyo garbage war" in May 1973, when garbage trucks were prevented from working. Furthermore, issues associated with the "urban heat island" phenomenon became evident.

## 2.2 South Korea

### 2.2.1 Overview of South Korea

South Korea's land area is 99,237 km$^2$, with a total number of inhabitants in 2015 of 50,620,000, making it the twenty-seventh most populated in country the world. In 2014, the urbanization rate was 82.4 % (NBSC 2015). The country is divided into a special city (Seoul), a metropolitan autonomous city municipality (Sejong), nine provinces (Gyeonggi Province, Gangwon Province, Chungbuk, South Chungcheong, Jeolla Province, Jeolla Namdo, North Gyeongsang Province, South Gyeongsang, and Jeju special self-governing province), and six metropolitan cities (Busan, Daegu, Incheon, Gwangju, Daejeon, and Ulsan). In the 1960s, South Korea's economy began to develop and continued to grow rapidly into the 1970s. South Korea's gross national income (GNI) per capita topped $20,000 for the first time in 2007, reaching $21,695. However, in 2008, the nation's GNI per capita fell to $19,296 under the impact of the international financial crisis. In 2009 it fell further to $17,193. However, in 2014, South Korea's economy grew by 6.2 % and the GNI per capita again exceeded $20,000 (NBSC 2015).

South Korea's economic potential is high, with the iron and steel, automotive, shipbuilding, electronics, and textile industries becoming the pillar industries, of which the shipbuilding and automotive industries are world renowned. Large enterprise groups occupy a very important position in the South Korean economy, with the value created by large enterprises such as Samsung, Hyundai Motor Co., SK, LG, and Korea Telecom representing more than 60 % of the national economy.

Seoul, the economic capital of South Korea, has a population of 10,140,000 (as of January 2014). Seoul is located in the northwest of South Korea in the River Han valley. It is the capital and the political, economic, technological, and cultural center of South Korea. The metropolitan area centered on Seoul, including most of the area in Incheon and Gyeonggi provinces, contains around 23 million inhabitants, making it the world's second-largest metropolitan area after Tokyo and hosting nearly half of the country's population. Seoul contributes 21 % of South Korea's GDP. The city is also the headquarters of many multinational companies and international banks (Fig. 2.4).

### 2.2.2 The Urbanization Process of South Korea

The urbanization of South Korea has undergone four stages: the stage of early urbanization, the stage of abnormal urbanization, the stage of rapid urbanization, and the stage of highly developed urbanization (Parka et al. 2011).

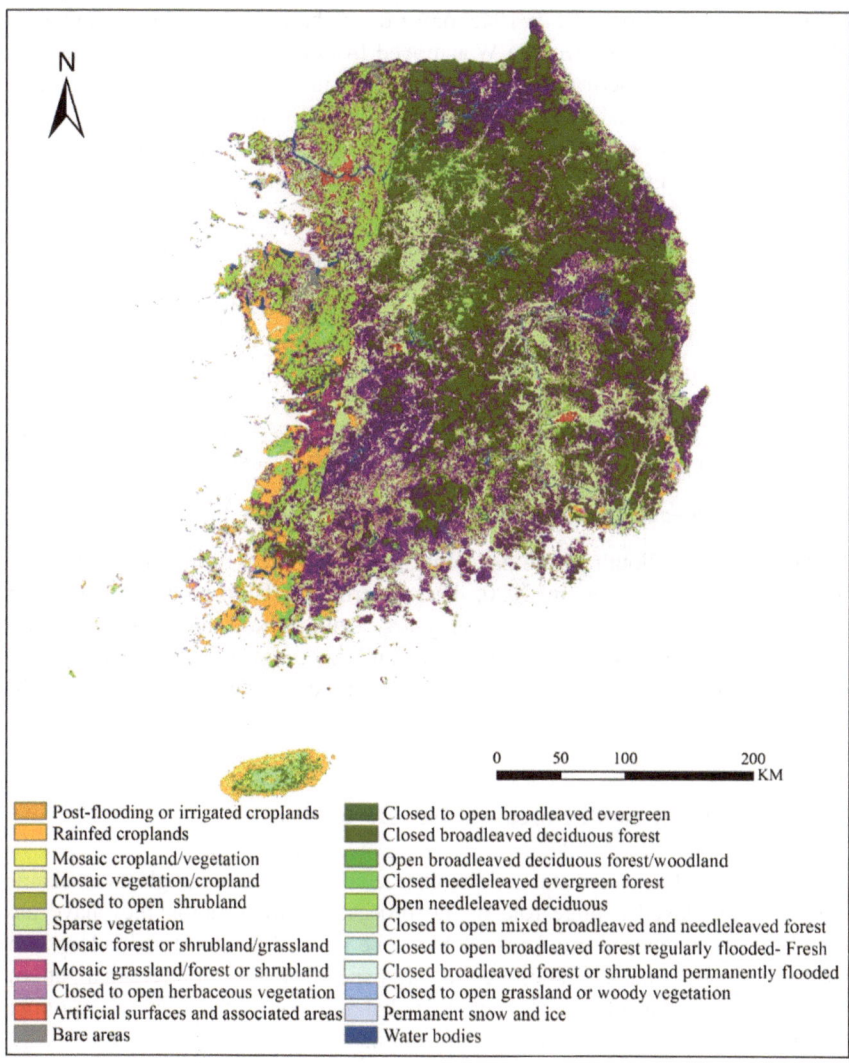

**Fig. 2.4**  Land use map of South Korea (*Data source* European Space Agency 2009)

### The first stage: early urbanization (1930s to Mid-1940s)

In this stage, South Korea was under colonial rule after Japan had occupied the Korean Peninsula in 1910. The percentage of urban population in the Korean Peninsula rose from 3.3 to 11.6 % by the mid-1940s.

### The second stage: abnormal urbanization (mid-1940s to early 1960s)

Urban population growth during this period was mainly due to migration caused by WWII. A large number of Koreans who had been exiled in China and Japan during

WWII returned to Korea from 1945 onwards. Subsequently, a large number of refugees generated by the Korean War moved from the middle part of the Korean Peninsula to the southern part, became engaged in cattle production around the cities, and settled down. The level of urbanization in South Korea rose to 28.3 % by 1960. This increase in the level of urbanization was caused by the removal of the colonial power and the migration of refugees after the war, both of which belonged to the abnormal stage of development.

**The third stage: rapid urbanization (early 1960s to late 1980s)**

During this period, the urbanization level rose from 28.3 % in 1960 to 74 % in 1985. Rapid industrialization accompanied this increase in urbanization. In 1968, the absolute population of South Korea's rural areas began to decrease, while the urban population exceeded the rural population for the first time in 1977. Although the population of Seoul in the late 1960s was less than 3 million, by 1988 it had increased to 10 million, making it one of the world's major cities.

**The fourth stage: highly developed urbanization (1990s to present)**

By the late 1980s, South Korea had achieved a high level of urbanization, which further increased to 82 % in 2000. The pattern of urbanization is likely to be adjusted accordingly during the twenty-first century to achieve a coordinated development of the country's economy, balance the uneven distribution of population, and solve a series of urban problems with the implementation of the capital relocation plan.

### 2.2.3  Features of South Korea's Urban Development

South Korean urbanization has unique characteristics, and although these characteristics have played a huge role in the process of urbanization they have also brought many associated problems unable to be ignored (as discussed below). The two most obvious characteristics are the "high speed" and the "preferential pattern" of urbanization. An additional feature is that the government and the political system guide the development of urbanization.

**High speed**

South Korea's urbanization has been extremely rapid. With the rapid pace of economic development and industrialization, many rural inhabitants have moved into cities. The government implemented an export-oriented development policy based on low wages and light industries. To increase exports, the government improved the production capacity of the existing plant, and subsequently the urban population also increased. New manufacturing companies have also established themselves in urban areas with good infrastructure and a willing labor force, adding to the pace of urban development. South Korea has experienced a very rapid pace of urbanization over the last 30 years and is now a highly urbanized country.

**The preferential pattern**

"Preferential pattern" means that urbanization did not proceed across the entire nation simultaneously, but first focused on a specific area. Korea obtained very high levels of economic growth in a short time through regarding economic efficiency as the first principle of measuring the location of economic activities, through establishing a centralized pattern of urbanization and industrialization (leading to benefits arising from proximity and scale), and through determining the pattern (spatial concentration) of urban development (Kim and Pauleit 2007). The preferential pattern of South Korea's urbanization created an obvious gap between big cities and small or medium-sized cities.

**The government and the political system play a guiding role in the process of urbanization**

South Korea has a highly centralized system of political power, which has been a key factor in promoting the highly concentrated development of the nation's cities. As a symbol of the system, the capital city of Seoul plays a prominent role in the national economic, political, and cultural life. It plays an outstanding role in the urban system and to some extent weakens the competitiveness of the non-capital cities and restricts the structure of the national city system. The Korean government and the political system have played an important role in the process of urbanization in South Korea, but have also brought many associated problems.

### 2.2.4  Challenges for South Korea's Urban Development

South Korean urbanization is dominated by large cities, which has allowed the positive economic effects of urban agglomeration to be gained. However, this dominance has also exacerbated the uneven development of different regions, resulting in a lack of dynamics in many local centers. As the dual social structure deepened and large numbers of people moved from rural areas to become industrial workers or the urban poor, the big cities expanded on the basis of the cheap labor and a certain material base. As urbanization accelerated, rural areas and undeveloped regions remained relatively backward, emphasizing and entrenching the dual structure of society.

**Big cities and the surrounding areas have a delicate relationship**

Big cities have been able to develop further because of the economic basis and political conditions, and during this process a spatial pattern has developed characterized by company headquarters being located in the big cities and branches of companies and plants being located in the surrounding smaller cities and regions. Through this process, the surplus value of local creation has in large part been transferred to the big cities (Jang and Kang 2015). In addition, the local urban divisions of factories face the prospect of being shut down first in an economic recession or a period of industrial structure adjustment. Moreover, the business class making the decisions is not local, which causes frequent labor disputes and problems with labor productivity.

**Urban infrastructure is weak with the increase urbanization**

With the influx of a large number of people into the cities, deficiencies in the urban infrastructure became apparent. As well as a lack of basic facilities, the poor quality of the infrastructure was also revealed to be a problem, including the occurrence of building or bridge collapses, damage to underground facilities, gas explosions, and other major accidents. Although these typify the problems that are known to occur in the process of urbanization, in South Korea the very rapid increase in the extent of urban areas aggravated the severity of the problems because the speed of the increase outstripped the ability to provide infrastructure of sufficient quality and consistency (Jeong and Ban 2014).

**Environmental pollution has become a problem with urbanization**

South Korea's rapid pace of urbanization has also increased levels of environmental pollution. Urban residents' living standards have improved over time, but the expansion and upgrade of material consumption associated with an improving urban lifestyle have increased the amount of waste produced, including atmospheric pollution, water pollution, and garbage/landfill.

**Urban policies have not always been suited to the development of cities**

Over the course of urbanization in South Korea, various policies were abandoned without implementation. Others were eliminated in the process of implementation, partly because of the speed of urbanization and the influence of the centralization of decision-making power, whereas policies conforming to these characteristics persisted. With the pace of economic development and urbanization, along with the problems associated with such development, transformations of urban policies became an inevitable occurrence.

## 2.3 Russia

### 2.3.1 Overview of Russia

The Russian Federation (in Russian, the "Российскаян Federati") is commonly known as Russia. It is a federal republic constitutional state consisting of 22 autonomous republics, 46 states, 9 krais, 4 autonomous regions, 1 autonomous prefecture, and 3 federal cities. Russia is located in northern Eurasia, occupying a position extending across the continents of Asia and Europe, and is the largest country in the world with a land area of $17,075,400$ km$^2$. Russia is an economic powerhouse and was the world's second-largest economy during the period of the Soviet Union, but its economy declined substantially after the collapse of the Soviet Union. However, the Russian economy developed rapidly after 2000 after selling a

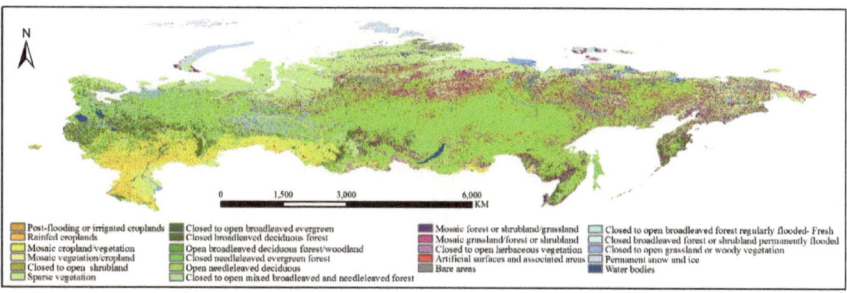

**Fig. 2.5** Land use map of Russia (*Data source* European Space Agency 2009)

large amount of natural resources. In 2006, the nation's total economic output topped that of its 1990 output within the Soviet Union. Russia's GDP reached $113.56 billion in 2007, ranking tenth in the world. In 2011, the gross GDP was $184.96 billion with an annual growth rate of about 4.3 %, and the GDP per capita was $12,939.

Russia has a solid industrial foundation covering all sectors. The main industries are machinery, steel, metallurgy, petroleum, natural gas, coal, forestry, and chemicals, and the wood and wood-processing industries are also fairly well developed. The industrial structure is somewhat unbalanced, favoring heavy industry over light industry, and the backwardness of civilian (i.e., non-military) industry has not fundamentally changed (Fig. 2.5).

### 2.3.2 The Urbanization of Russia

The development of cities in Russia occurred later compared with Western developed countries. Urbanization in Russia has maintained the principal characteristics of the development of cities during the Soviet Union period, including the rapid pace of development, a strong national system, and the lack of a societal middle class. Russia's cities have developed on the basis of the urbanization that took place during the period of the Soviet Union. Russia's urbanization resembled the trend of the national economy after the collapse of the Soviet Union, which fluctuated widely.

It can be seen from the change in Russia's urbanization rate (Fig. 2.6) and the annual average growth rate of urbanization (Fig. 2.7) that the urbanization process has gone through the initial stage and subsequent development stages and now appears to have reached a plateau of a high level of urbanization. Historically, Russia was a largely agricultural country, with a total population of 67 million in 1897, of which only about 15 % lived in cities. By October 1917, Russia's total population had increased to 91 million, but the proportions of urban and rural population had not fundamentally changed.

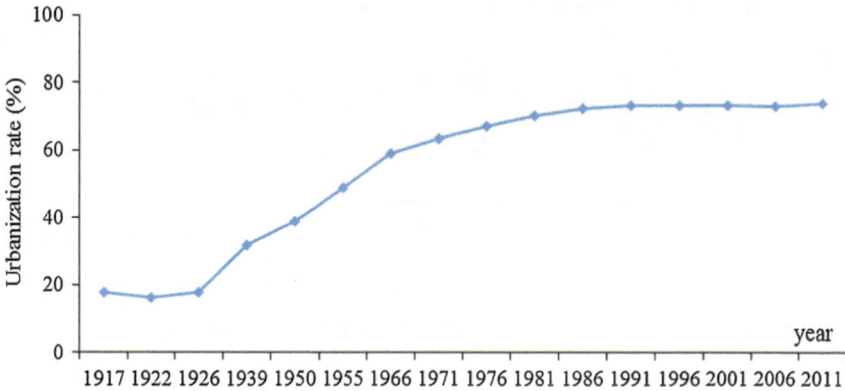

**Fig. 2.6** Change in the urbanization rate of Russia (*Data source* World Bank)

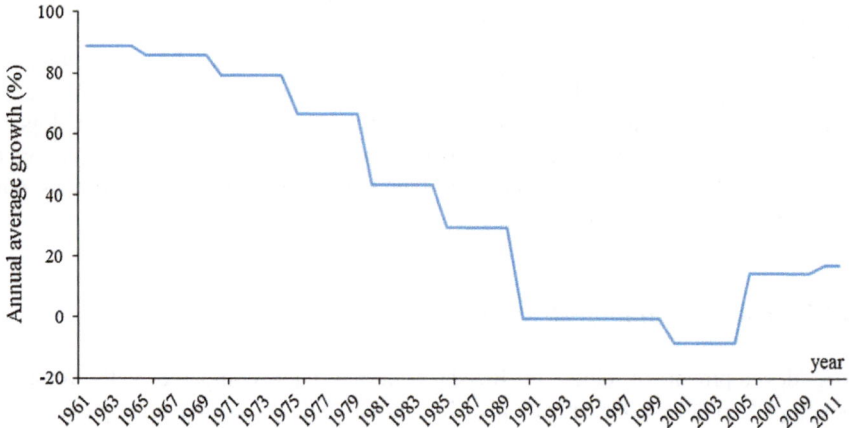

**Fig. 2.7** Annual average growth rate of Russian urbanization (*Data source* World Bank)

In 1925, the Communist Party of the Soviet Union passed the socialist industrialization policy in the 14th Congress and decided to transform the Soviet Union from an agricultural country to an industrial country in a series of steps. This marked the beginning of the urbanization of the former Soviet Union, characterized by rapid rural–urban migration. The Soviet population, including Russian, soared to 108 million in 1939, with the urbanization rate increasing to 33 % over the same period.

WWII (1939–1945) brought a significant reduction in the population of the Soviet Union. However, the previous growth trend then continued, and by 1959 the Soviet population had increased to 120 million with the urbanization rate increasing to 52 %. Subsequently, the Soviet Union experienced continuous population increase and urban development in the 1960s, 1970s, and 1980s, with the population increasing to 140 million by the end of 1989 and the urbanization rate increasing to 73 %. This period could be regarded as the golden age of rapid growth of both the Russian population and its urbanization.

The collapse of the Soviet Union brought the golden age to a halt in the early 1990s. Although the percentages of urban (72 %) and rural (28 %) population did not change significantly, the Russian urban population and the total population declined between 1991 and 2005.

**The five stages of the urbanization of Russia**

The urbanization of Russia can be divided into five stages on the basis of the above discussion.

**The first stage** (from 1917 to 1926): Russia was still a largely agricultural country before the establishment of the Soviet Union. The total population increased to 91 million and the urbanization rate increased to 18 % by 1917.

**The second stage** (from 1926 to 1939): The number of cities and towns increased rapidly, the population soared to 108 million, and the urbanization rate increased to 33 %.

**The third stage** (from 1940 to the collapse of the Soviet Union in 1991): Both the overall population and the urbanization rate increased rapidly, particularly from the 1960s onwards in the golden period of growth. By the end of 1991, the urbanization rate had increased to 73 %.

**The fourth stage** (from 1991 to 2005): Affected by the downfall of the Soviet Union, the process of urbanization in Russia stagnated and the phenomenon of counter-urbanization appeared.

**The fifth stage** (from 2006 to present): Population migration to the cities restarted, reversing the trend of the fourth stage.

### 2.3.3  Characteristics of Russia's Urbanization

**The structure of most urban functions is single**

According to Russian experts, there are two standards to determine whether the city's industrial structure is simple. The first is where a single enterprise or the same industry has created more than half of the city's industrial output value or service industry output value. The second is where a single company engages more than a quarter of the working population. The single structure of a city has both advantages and disadvantages (Wu et al. 2016). If the product or products concerned have international competitiveness, it is easier to obtain foreign investment and the support of national policies, so the host city is able to be further developed. However, many cities struggle because of constraints on energy supply, which provides a major constraint on production and competitiveness.

## The development of urban agglomeration is slow

Russia has not yet formed a series of cities in the real sense. In the future, such a series is likely to form along two paths, one along the Moscow–St. Petersburg railway and highway, and another along the Moscow–Nizhny Novgorod highway.

## The level of urbanization has increased

The development of Russian cities has a strong historical inheritance, with most cities being transformed directly by rural–urban migration or by the reform of administrative divisions when many villages became cities or towns virtually overnight in order to meet the needs of rapid industrialization.

## The spatial variation in the level of urbanization is significant

The distribution of cities is very uneven across various regions of Russia, and the most obvious demonstration is that the cities in the west outnumber those in the east. From a national perspective, more than 25 % of the cities are concentrated in the western core area covering less than 4 % of the total land area, whereas fewer than 5 % of the cities in the country are distributed in the Far East with more than 33 % of the total land area. On the whole, the geographic distribution of Russian cities is consistent with the distribution of population.

### 2.3.4 Challenges for Russia's Urban Development

The development of urban areas is subject to the strict planning of the state, meaning that top-down planning dominates the urbanization model. The urbanization of the Soviet Union was achieved mainly by relying on the approach of deployment and dispatch of the labor force by the government, along with the combination of implementing applications for citizenship certificates and the declaration of the hukou system, which aimed to guarantee government control of the whole process of city development. Soviet urbanization showed distinct characteristics resulting from the particularities of the economic base, the political system, geographic conditions, history, and culture. The process was characterized by the gradual promotion of urbanization in the national plan, the promotion of urbanization relying on high-speed industrialization, urbanization lagging behind industrialization, and the unbalanced development of urbanization (Wang et al. 2016b).

At first, urbanization under the Soviet system occurred rapidly. The urbanization rate increased from 18 % in 1917 to 65 % in 1985. In those 68 years, the Soviet population increased from 136.0 million to 273.8 million while the urban population increased from 29.0 million to 177.5 million. Also, although the Soviet government formulated policies to control the development of big cities with respect to the distribution of urban population, the population growth rate of the big cities was higher than that of the small and medium-sized cities. According to the 1979 census, the population of metropolitan areas in the Soviet Union accounted for

12.6 % in the total population, 20.3 % of the urban population, and 33.5 % of the population of the big cities. Furthermore, the rapid development of urbanization was the result of the implementation of the rapid industrialization strategy. The rapid industrialization of the Soviet Union was dominated by heavy industry, but the gap could be filled only by moving a large number of migrant workers from rural areas as well as utilizing rich natural resources.

However, there were many problems that accompanied the urbanization of the Soviet Union. These included a serious urban housing shortage, weak urban management, stunted development of rural areas and agriculture, and the weak development of light and tertiary industries.

## 2.4 India

### 2.4.1 Overview of India

India is the largest country by land area in the South Asian subcontinent and seventh-largest in the world. As one of the oldest civilizations, India has a brilliant diversity and rich cultural heritage. India is the world's second-most-populous country, with a population of 1.27 billion in 2014, accounting for nearly one-fifth of the world's population (NBSC 2015) (Fig. 2.8).

### 2.4.2 Urbanization Process of India

**India began urbanization in the late nineteenth century and the development process can be divided into three stages**

**First stage, 1901–1921**. In the first stage, the development of urbanization was extremely slow and the annual growth rate of the urban population was only 0.79 %. In 1901, India's urbanization rate was 10.8 %; as a comparison, China is still below this level.

**Second stage, 1921–1951**. Urbanization developed rapidly with the annual growth rate of the urban population being 3.52 %. India became independent of British rule in 1947 and entered the take-off stage of urbanization. India's modern industrial production accounted for 17.1 % of national income by 1949, with the number of industrial workers reaching 13 million. In 1950, India's urban population was 17.3 % of the total.

**Third stage, 1951–present**. Urbanization developed rapidly, with the annual growth rate of the urban population increasing from 2.35 % in the 1950s to 3.8 % in the 1970s, and reaching 3.89 % in 1981. At the beginning of the 1990s, an international payments crisis forced the Indian government to carry out economic reforms, enabling the nation's economy to develop in the directions of "liberalization" and "globalization". In 1990, India's urbanization rate was 25.7 %, and the

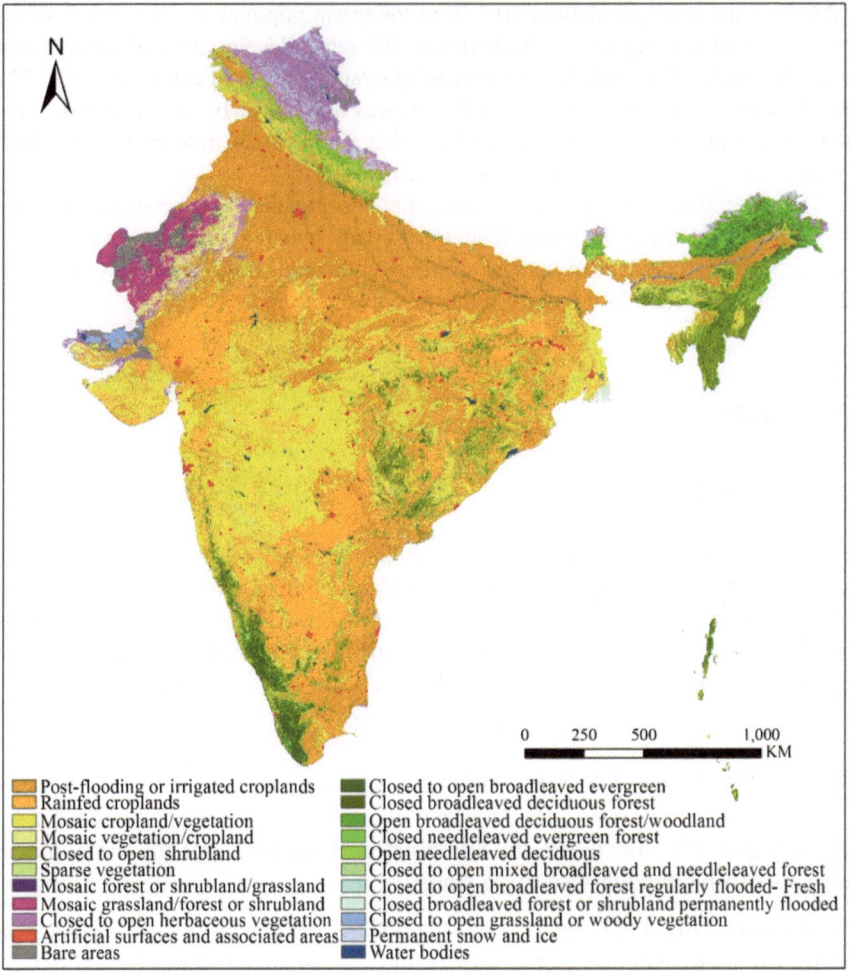

**Fig. 2.8** Land use map of India (*Data source* European Space Agency 2009)

average annual economic growth rate was more than 7 % from 1992 to 1994, although was somewhat less (around 5 %) in other years, making it one of the world's fastest-growing economies.

### 2.4.3 Characteristics of India's Urbanization

**The urbanization rate was lower than the average for low-income countries**

From the perspective of an international comparison, the speed of urbanization in India was slower compared with most other countries (Swerts et al. 2014). Before the 1990s, India's urbanization rate was higher than the average of low-income countries but since the 1990s the rate has been lower than the average of low-income countries. In 2005, the average urbanization level in low-income countries (India included) was 29.95, 1.26 % higher than that in India. If India were not included in these figures, the gap would be larger (2.35 % higher). Urbanization in India was also characterized by an increase in the concentration of population in large cities (Dewan and Yamaguchi 2009). In 1960, the urban population in Indian cities with more than one million people accounted for 37.0 % of the total urban population, reaching 37.8 % in 1990 and 40.5 % in 2005.

**Service industries grew quickly in the process of urbanization and economic growth**

India's economic growth rate was low for a long time, but accelerated after the reforms of the 1990s. From 1961 to 1990, the average annual GDP growth rate was 4.2 %, increasing to 6.0 % from 1991 to 2005. The predominant feature of India's economic growth was the rapid development of the service industry sector. From 1961 to 1991, the average annual growth rate of service industries was 5.3 %, increasing to 7.9 % between 1991 and 2005, which was in stark contrast to secondary industries and agriculture. From 1961 to 1990, the average annual growth rate of secondary industries was 5.5 %, increasing to 6.2 % between 1991 and 2005. The annual growth rate of agriculture in these two periods was 2.4 % and 2.6 %, respectively. Correspondingly, the contribution of service industries to the GDP increased from 34.3 % in 1961 to 42.1 % in 1991 and 54.4 % in 2005. In contrast, the percentage contributed by secondary industries was 27.3 % in 2005, having increased by less than 1 % since 1991 (Sharma and Chandraskekhar 2014).

**The natural increase of the urban population was the main factor increasing the urbanization rate**

The natural growth of the urban population was the main factor in India's urban population growth, which was quite different from other developing countries (Pandey and Seto 2015). The three main drivers of a nation's urban population growth are natural population growth in cities and towns, the migration of people from rural areas, and the redefinition of urban areas. According to India's census data in 1991, about 41 % of the increment in India's urban population between 1971 and 1981 was caused by natural population growth in cities and towns, and about 36 % by migration from rural areas and the redefinition of urban areas. However, for the period 1981–1991, the two ratios were 60 and 22 %, respectively.

**The relationship between rural and urban areas in India has space to improve in the process of urbanization**

The driving mechanism of Indian urban population growth was the push of rural poverty rather than the pull of urban prosperity. Although this migration of people from rural areas was not such a large contributor to urban population growth, the dynamic mechanism of migration (rural poverty) was different from that experienced in most other countries (Das et al. 2015). One consequence of this dynamic mechanism was a rise in the number of people living in slums in Indian cities, as the lack of urban prosperity meant that cities were not ready to receive the additional influx of people (Moghadam and Helbich 2013). Although India implemented land reform after independence in 1948, it was not entirely successful, meaning that large tracts of land were neither farmed nor used for residential purposes.

The gap in the standard of living between urban and rural areas has not significantly narrowed. In fact, in the last 20 years, the urban–rural income gap in India has widened sharply. In 2008, based on survey data of representative families, The World Development Report found that the urban-to-rural median income ratio increased over the period 1989–1999. In addition, the gap between urban and rural poverty has also widened in India, even though there has been an overall decline in the poverty rate. The poverty rate in urban and rural areas was 32.4 and 37.3 %, respectively, in a 1993–1994 survey, and was 24.7 and 30.2 %, respectively, in a 1999–2000 survey.

### 2.4.4  Challenges for India's Urban Development

#### Urban infrastructure is weak

The piped water coverage in Indian cities is only 74 %. The current urban water supply per capita per day is 105, 45 L less than the international standard of 150 L and much less than the highest international standards of 220 L. According to the current trend in urban population requirements, it is estimated that by 2030 the daily water needs of Indian cities will be 189 billion liters, while the cities' daily water supply can currently provide only 95 billion liters.

#### Many poor people live in urban slums

According to India's 2001 census data, 42.6 million people live in slums in cities with populations of more than 50,000, accounting for 22.6 % of the total urban population. Moreover, while the number of people living in slums has grown rapidly, the number of slums has not changed, leading to a marked increase in the population density in the slums (Cohen 2006).

#### There is a low quality of city management and low efficiency in city operations

In major cities such as Delhi and Mumbai, all kinds of vehicles can be found on over-congested, poor-quality urban roads. Vehicles are seriously overloaded. The general congestion and poor infrastructure hampers the efficiency of urban operations such as rubbish removal and maintenance services.

#### Social security, health care, and compulsory education are poor

Children living in slums are able to get free education and subsidies for school uniforms and food in the compulsory education stage. However, most public services for social security and health care are not able to meet the needs of such large numbers of people because of limited supply capacity.

## 2.5 Indonesia

### 2.5.1 Overview of Indonesia

Indonesia is the world's fourth-most-populous nation, and its inhabitants are spread over 13,000 islands. National transportation networks are poorly developed, and the level of economic development places the nation in the middle of developing countries. The per capita gross national product (GNP) in 1995 was $980 and reached $1280 by 1999. Education provision has developed rapidly and health services have improved. Owing to the implementation of a family-planning policy, the population growth rate has dropped to slightly less than 2 % since the 1980s.

Indonesia is the largest economy of the Association of Southeast Asian Nations (ASEAN), with the agricultural, industrial, and service industries all playing important roles in the national economy. Indonesia produces the second-largest amount of economic crops, such as palm oil, rubber, and pepper, in the ASEAN region (Wang et al. 2016a). The direction of industrial development is to strengthen export-oriented manufacturing, and the rate of industrial growth has been higher than the rate of economic growth in recent years. Indonesia has a wealth of mineral resources, including oil, natural gas, coal, nickel, lead, copper, gold, silver, chromium, bauxite, sulfur, and kaolin. In addition, quantities of manganese, uranium, feldspar, marble, granite, quartz sand, clay, and dolomite are produced.

Indonesia's forest cover rate is 67.8 %, with a forested area of 120 million hectares, of which permanent forest covers 112 million hectares and convertible forest covers 8 million hectares. Indonesia is rich in a variety of rare tropical species, such as ironwood, sandalwood, and ebony, all of which are world renowned. Food crops are the foundation of Indonesia's planting industry, with rice being the staple food and corn, cassava, and beans being the alternative products. Indonesia is the largest producer of beans in Southeast Asia, although the unit output of beans is low, pointing to low efficiency in some agricultural activities (Fig. 2.9).

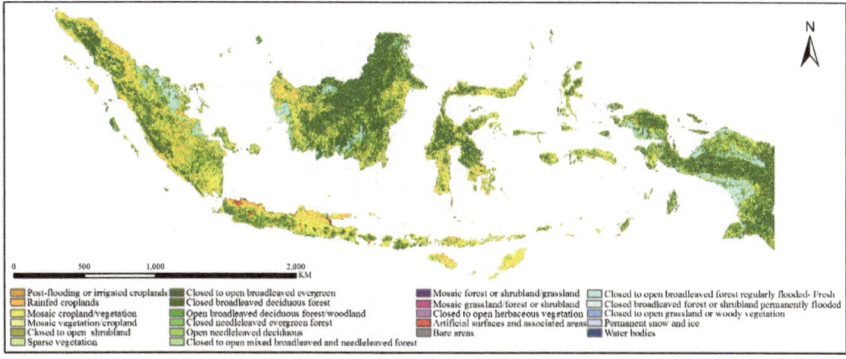

**Fig. 2.9** Land use map of Indonesia (*Data source* European Space Agency 2009)

### 2.5.2    The Urbanization Process of Indonesia

Like most developing countries in Asia, Indonesia is a country with low levels of urbanization and industrialization. Its urbanization process can be divided into four stages.

#### The first stage (before 1970): the stage of slow urbanization

Indonesia had experienced 300 years of colonial rule before achieving independence in the late 1940s. Indonesia had only a small number of cities throughout the colonial period. The largest city, Jakarta, had only 500,000 people, and only 7 % of the country's population lived in urban areas. A socialist regime ruled for almost 20 years immediately after independence. In that period, Indonesia on the one hand had to treat the political and economic trauma caused by the long-term colonial rule and revolutionary period, and on the other hand it had to strengthen the political institutions of the country and its various regions. Dramatic institutional change and the concentration of political governance stimulated the rapid growth of cities. A number of new cities were formed, with the total number of cities in 1965 exceeding 20, the percentage of the population living in cities reaching 15 %, and the urban population reaching 15 million. However, most of the growth of urban population became concentrated in a few large cities; for example, Jakarta's population reached more than 3 million and the population of Surabaya and Ban Dung both exceeded 1 million. In this first stage, Indonesia's urbanization process was paced and the development was not balanced, with urbanization being characterized mainly by the rise of a few large cities.

#### The second stage (the 1970s): the stage of rapid urbanization

The 1970s are known in Indonesia as the best 10 years of development, that is, "the golden 10 years," during which the economy and society of Indonesia changed dramatically. The Suharto regime implemented political reform and took a series of measures to promote the development of the domestic economy, including the implementation of import substitution policies to foster and protect the development of domestic industries, the implementation of an agricultural green revolution to achieve self-sufficiency in rice and other food, and the encouragement of foreign investment to exploit the rich domestic resources, which together saw the nation's economy achieve rapid growth in the 1970s. Correspondingly, urbanization also experienced a stage of rapid development. According to official statistics, Indonesia's urbanization rate reached 22 % by the end of the 1970s, with 31 million people living in urban areas, including 9 cities with populations of more than 500,000, 13 cities of 200,000–500,000, and 42 cities of 100,000–200,000. However, the growth of the urban population in this period was concentrated mainly on the island of Java as a result of the imbalance of the historical development process and different resource conditions. At that time, the urban population of this island accounted for 70 % of the total urban population of the country, with the metropolitan areas of Jakarta, Dong, and Suragbaya accounting for 40 % of the population of Java.

### The third stage (the 1980s): the stage of extraordinary urbanization

The policies and actions that had been used to stimulate economic growth in the 1970s in Indonesia met challenges in the 1980s. The reallocation of the international market had a major impact on Indonesia's industries, with three major aspects being of most influence. The first was the opening of Southeast Asia, especially China, to the outside world, which meant that other nations in the region attracted large amounts of international capital, placing competitive pressure of foreign capital utilization onto Indonesia. The second was a fall in the prices of exported raw materials and other products that had influenced Indonesia's economic boom. The third was that domestic agriculture, especially rice production, had already attained very high levels, and it was therefore difficult to achieve greater levels of production. All these effects reduced the government's financial revenue, and the national economic growth rate dropped to 2.5 % per annum.

Normally, Indonesia's urban development would have been expected to slow in pace during this period. However, urbanization continued to develop rapidly in the first half of the 1980s because of the inertia of the economic boom in the 1970s and the concentration of rural land in the 1980s. Indonesia's urbanization rate had reached 31 % by the 1990s according to statistics, an increase of 9 % compared with 1980. The increase in urban population was concentrated mainly on the island of Java. Java absorbed 70 % of the new urban population in this period, making the overall level of urbanization on the island increase to 36, 5 % higher than the national average. The urban corridor dominated by Java experienced a period of exceptional population growth, driven mainly by the large and growing urban population base. For example, the urban population growth rate of Jia Bata Beck city reached 5.8 % per year in the 1980s and it subsequently became a megacity of 13 million people (Firman 2009).

### The fourth stage (since the 1990s): period of urbanization adjustment

Entering the late 1980s, the rapid urbanization of Indonesia began to encounter some resistance and exposed some problems. First, the urban population was too concentrated in Java's major metropolitan areas, resulting in over-crowded conditions and a lag in urban infrastructure, including transportation and utilities. Second, the urban population was very sparsely distributed in the vast unexplored regions of Indonesia's outer islands, resulting in a marked imbalance in the level of regional development and triggering a number of political and/or ethnic instability factors.

As a result of these problems, Indonesia's urbanization entered a period of adjustment in the 1990s. The main features of this period were as follows. The government intended to take measures to adjust various aspects of the development of the nation's urban areas, but the adjustment measures failed to work properly because of the previous accumulation of problems. During this period, the government tried to encourage the migration of population from within Java Island (Javanese) to the eastern islands. Unfortunately, this policy caused population movements only within the island of Java itself, especially the flow of the rural population into urban areas, because of constraints on movement and the lagged development of other islands. The government also adopted more favorable policies to encourage foreign investment to move into other islands, but this investment was preferentially attracted to capital-intensive and non-labor-intensive industries because of the restriction of resources and competition in the international market. Therefore, the effort to develop urbanization by encouraging it to follow industrial development did not work well. As a result, the development of Indonesian urbanization remained largely concentrated in the island of Java, while the outer islands developed only slowly in the 1990s.

### 2.5.3 Characteristics of Indonesia's Urbanization

**Excessive concentration and imbalance in the development of urbanization**

Urban development in Indonesia continues to be concentrated mainly in the corridor region of Java Island along Sumen Tana–Jakarta–Bali–Ban Tong. According to statistics, about 80 % of the 45 million urban population lives in this corridor region, almost all of the large cities (>500,000 people) are concentrated in this region, and about 80 % of all the cities are found in this region. Because of this, the average urbanization rate in this region is as high as 40 % compared with the national urbanization rate of 32 %. In contrast, cities are few and the urban population is low in the majority of the outer islands (Murakamia et al. 2005).

**The industrial orientation of urbanization, especially the role of foreign investment industries, is very obvious**

The distribution of industry, especially the concentration of industries related to foreign investment, has played an important role in guiding the development of urbanization in Indonesia. No matter whether it was the inflow of capital and development of industry by the British and Dutch in the early colonial era, or the inflow of capital from Japan, Korean, and Germany in post-independence days, the Java–Bali region was chosen as the industrial base, thereby stimulating the development of local cities and attracting large numbers of workers into employment in those cities. From 1967 to 1982, 67 % of foreign direct investment was concentrated in the Java–Bali region, with foreign investment, industrial concentration, and urbanization going hand-in-hand to produce the particular pattern of development seen. According to World Bank statistics, the per capita GDP of the outer islands was higher than that of Java Island in 1976, but by 1987 the per capita GDP of Java Island was 25 % higher than that of the outer islands, reflecting the concentration of industries in Java (Goldblum and Wong 2000).

**The government's inaction caused a concentrated pattern of urbanization and industry to develop**

Indonesia has generally not had a clear industrial layout policy, at least until the 1980s, which meant that the distribution of industry never took account of the needs of the country to have a balanced development of both its economy and its urbanization. State-owned enterprises, private enterprises, and foreign capital enterprises all chose to develop in the Java–Bali area, which had a significant economic advantage, meaning that 80 % of all large and medium-sized industrial projects in Indonesia were concentrated in this area by 1990. After entering the 1990s, the government tried to encourage industries, including those related to foreign investment, to develop in the outer islands, as well as to attract people to move to those islands. However, the development of the outer islands was limited to capital-intensive industries, including the oil industry, because of logistical barriers and other difficulties that could not be overcome by government incentives. The movement of population was limited to within the island of Java because there was insufficient incentive for people to migrate to the outer islands. Statistics show that 75 % of urban dwellers lived on the island of Java in the 1980s and 1990s. The above points indicate the ineffectiveness of the Indonesian government in directing the pattern of urbanization, possibly because of a lack of real motivation to do so.

**The main pull force of urbanization: non-traditional, previously rural-based industries are now concentrated in cities**

The pull of cities in the process of urbanization in Indonesia is explained mainly in terms of their attraction of non-traditional industries rather than in the expectations of better urban employment and income. According to research, many non-traditional industries related to the livelihoods of rural residents, such as the processing industry, handicraft industry, and textile industry, were distributed in rural, not urban, areas before the stage of very rapid urbanization in Indonesia in the 1970s and 1980s. However, these non-traditional industries, which were originally distributed in rural areas, became rapidly concentrated in urban areas in the 1970s and 1980s as a result of national economic policies and general trends, and established themselves as informal industries of the cities. Statistics show that these informal industries contributed 25 % of cities' industrial growth and 30 % of urban employment growth in the same period, and were the main factor in attracting and absorbing rural workers into urban areas.

**The main push force from rural areas was the concentration of rural land ownership forcing farmers to move to cities**

Rural land ownership changed dramatically after the implementation of the capitalist system. The free sale of land and private ownership allowed rural land to become concentrated in the hands of a few people, so peasants with no land or only small plots of land began to appear in rural areas. In addition to the loss of comparative advantage of agricultural products, farmers with no land were unable to find a livelihood in the countryside, forcing them to move to the cities and into non-traditional, previously rural-based industries. The population pushed from rural areas accounted for around 10 % of cities' population increment each year.

### 2.5.4    Challenges for Indonesia's Urban Development

It is clear that the process of urbanization in Indonesia shares some common features and problems with other developing countries, as well as having its own particular unique issues.

#### Urbanization development is extremely concentrated

The urbanization rate in Indonesia reached 32 % in the 1990s and the total urban population is about 45 million people, with about 80 % of the urban population being concentrated in the Java–Bali region. All major cities (population >500,000) and more than 80 % of all cities are concentrated in this region, and there still remains great potential for further concentration. The reasons for this centralized distribution of population are the historical processes and structures and the highly imbalanced nature of economic development. Around 92 % of the total population is concentrated in Java, Sumatra Tana, Clement Tor, and Suna Weixi. In 1990, Java Island accounted for only 7 % of the area of the country, but contained 60 % of the national population.

Indonesia's per capita gross national product is below the average level of developing countries. This lack of economic strength has not allowed Indonesia to develop sufficient infrastructure and industry in the outer islands. World Bank statistics point to the situation of uneven development of cities and towns in Indonesia (and of related industries), as well as the consequences of this imbalance. In 1985, the densely populated cities of Java Island contained 70 % of all job opportunities in the manufacturing sector (which indicates the high degree of concentration of manufacturing industry in Java Island compared with other parts of Indonesia), while the broad regions including the outer islands provided only 6 % of manufacturing jobs. Also, 72 % of Indonesia's national construction projects are located on Java, indicating a reinforcement of the imbalance in economic development (Fahmi et al. 2014).

#### Increased urban environmental hazards and associated costs

In Indonesia, the large cities (such as Jakarta) are estimated to discharge 30 % of their domestic waste (including human waste) into rivers and canals because of the lack of sewage facilities. This has resulted in serious water insecurity, and the spread of illnesses such as diarrhea, typhoid, and cholera, which are the main cause of the high infant mortality rate (about 74 per thousand in the 1980s). In addition, the World Bank estimates that if there are no strict measures put in place by 2020, then the following pollution levels will occur with respect to present levels: the organic nitrogen in water will increase by 10 times; air-suspended particles will increase by 14 times; and harmful metals (e.g., mercury and lead) will increase by 20 times. The combined effect of all of these problems will constrain the economic development of Indonesia's cities. In 1990, Indonesia disbursed up to $500 million in environmental expenditure according to World Bank statistics in an effort to ameliorate the damage to health and the environment caused by pollution. In addition, in the 1990s, $1 billion each year was spent on improving the urban environmental infrastructure. A similarly serious problem is that the cities have become too concentrated, which has led to severe water shortages in areas around the big cities. The water that many people have relied on for survival for decades has largely gone, and some of the most suitable production areas for rice have been unable to maintain production. More than 8 % of the farmland is lost each year in the Yaba Taback region alone.

**A serious shortage of urban housing**

The quality of Indonesian housing is generally low compared with that of other developing countries. Indonesia's average urban housing area per capita is less than 10 m$^2$, and the distribution is very unequal, with about half of urban inhabitants having less than 3 m$^2$. A substantial number of people had no house in which to live during the rapid development of urban areas in the 1970s. A national housing improvement program was started in the mid-1970s, in which the government developed a series of cheap and practical houses containing basic facilities. At present, low-income families can expect to live in a house of 15–21 m$^2$, and middle-income families in a house of 45–70 m$^2$. It is estimated that about 300,000 new houses are needed in Indonesia every year to meet the needs of urban development. However, the housing provided by the Indonesian government and the private sector can meet only a small part (about 30 %) of the annual housing needs.

**The problems continue to accumulate**

Ethnic divisions, social instability, urban over-crowding, and excessive concentration of urbanization have made the development of Indonesia appear extremely unbalanced. The high rate of development hides and continues to exacerbate serious socio-political problems, including inter-ethnic discord between regions and social and political instability.

## 2.6 Israel

### 2.6.1 Overview of Israel

Israel is located in the southeast part of the Mediterranean, bordering Lebanon to the north, Syria and Jordan to the east, and Egypt to the southwest. Israel was declared an independent state in 1948. The country has a population of 7.7 million, of which 76 % are Jews (mostly Ashkenazi Jews), 20 % Arabs, and 4 % "other" races as of May 2011, according to data from Israel's Population Statistics Department. Of the Jews, 68 % were born in Israel, usually the second or third generation of Israelis, and 22 % of the remaining 32 % of foreign-born are from Europe, and 10 % from Asia and Africa, including the Arab world. Israeli scientists' contributions in genetics, computer science, optics, engineering, and other technology industries are recognized as being outstanding. The nation's most well-known research and development industry is the military technology industry, followed by agriculture, physics, and medicine.

### 2.6.2  The Urbanization Process of Israel

Israel is one of the most urbanized countries in the Middle East, with the level of urbanization reaching 90.8 % in 1995 and increasing slightly to 91.7 % in 2005. The process of urbanization in Israel can be divided into two stages.

**The first stage: a steady and then rapid development of urbanization (1870–1948)**

In the 1870s, the region that was to become the modern State of Israel had 10 cities with a total urban population of 120,000, accounting for 25 % of the total population of the region. Owing to the slow development of industry and the small agricultural hinterland, the scale of these cities was small. Urbanization developed more rapidly in the 1880s, driven by an influx of large numbers of immigrants, which increased the urbanization rate as well as the number of cities. In 1948, Tel Aviv's population soared to 248,000, becoming the only city in Israel with more than 100,000 people. The number of cities containing more than 10,000 inhabitants increased to nine, with the urbanization rate reaching 50 %. Despite the rapid development of urbanization, the total urban population at the end of this stage was still relatively low and the number of cities was small. In addition, the geographic distribution of cities was unbalanced, with more cities along the coast than inland.

**The second stage: sustained, rapid urbanization promoted mainly by the government (1948–present)**

After the founding of the State of Israel, migration of Jews into the country increased dramatically, providing the labor force and a growing domestic market for the rapid industrialization of the nation. Under the dual drive of industrialization and immigrants, the urbanization of Israel also accelerated. Between 1950 and 2000, the population of the nation increased from 1.37 to 6.28 million, with the number of foreign immigrants reaching 3 million and the urban population increasing dramatically from 813,000 to 5.74 million. The level of urbanization in Israel increased substantially from 60 % in 1948 to 84.2 % in 1970 and 91.2 % in 2000. The number of cities also increased rapidly in response to immigration and economic development. Between 1948 and 1984, Israel established 36 new cities, and between 1985 and 1995, it established more than 130 new cities (including satellite cities). During this drive towards greater levels of urbanization, modern secondary industries developed and grew strongly, as did tertiary (service) industries. The importance of secondary and tertiary industries to the national economy continued to increase, creating a large number of jobs. The scale of urban development also expanded, as the Israeli government started to address the geographic imbalance of urbanization (whereby cities tended to be concentrated in coastal and central regions) by formulating a number of plans to alter the regional layout of urban areas.

The population distribution in various parts of Israel was not balanced before the founding of the state. In 1948, about 75 % of the population was concentrated in the area of Haifa and Tel Aviv, which accounted for only 11 % of the nation's land, and about 80 % of the cities were located in the coastal zone. In contrast, only 8.4 % of the total population lived in Gailiesai and the Negev District, which together account for 70 % of the overall land area. In 1961, more than 50 % of the population lived in the central area and Tel Aviv, with 20 % of the population being distributed in either the north or south of the country. The population of the central area and Tel Aviv dropped to 45 % of the total by 1983. In the 1960s, the Israeli government focused on the development of the southern and northern parts of the country, especially directing industrial investment to those parts and also providing them with cheap land, tax incentives, export subsidies, and other policies or actions. In addition, the government achieved a more balanced geographic distribution of the domestic population through the rational planning of industrial layout.

### 2.6.3 Characteristics of Israel's Urbanization

**Urbanization is closely linked to national security strategies**

Because of Israel's special national conditions, its urbanization has been placed into the national security strategy framework, and this is the most prominent feature of Israeli urbanization. For national security reasons, the Israeli government decided to build new towns in both the south and the north of the country for resettlement. Since 1967, Israel has focused on the development of the newly occupied West Bank and the Gaza Strip.

**International migration has boosted the development of urbanization in Israel**

The increase in the urban populations of most countries relies mainly on natural population growth and internal migration. Israel, however, has relied more on international migration. According to statistics, the first peak of international migration occurred between 1948 and 1952, shortly after the founding of the State of Israel. Around 680,000 people migrated to Israel during those four years, and the country's Jewish population more than doubled. International migration has continued to be a characteristic of Israeli urbanization. Since the founding of the state, immigrants moving to the country have come mainly from Eastern Europe, Western Europe, and Islamic countries. Jews who immigrated to Israel from Eastern European countries were mainly small business owners and those engaged in service industries, and tended to settle in the smaller cities. Well-educated Israeli Jews who emigrated from Western Europe worked mainly in administration, commerce, and banking, and settled in the larger cities. These immigrants not only provided Israel with a source of labor, but also played an important role in boosting the establishment of new towns.

**The establishment of new towns has been an important part of the pattern of urbanization in Israel**

Another feature of Israeli urban development has been the establishment of new cities. On the one hand, these new cities have been able to release the pressure of immigrants by absorbing them, but on the other hand the process has added to the problem of the domestic population distribution being too concentrated. From 1949 to the early 1980s, the Israeli government built more than 30 new towns and settled about 400,000 residents in them. The number of newly built cities has varied over time. From 1948 to 1951, 22 such cities were established, and the population of the people settled in them was 126,081, accounting for about 9 % of the country's Jewish population. From 1952 to 1957, Israel built 10 new towns, hosting a population of 254,095, accounting for 14.1 % of the country's Jewish population.

### 2.6.4 Challenges for Israel's Urban Development

**The overall population is increasing along with a rising urbanization rate**

With the establishment of the State of Israel, a large number of immigrants arrived in the country. From 1950 to 2005, the population in Israel grew from 1.4 to 6.9 million, and the urban population also increased dramatically over the same period from 813,000 to 6.132 million, increasing by 7.5 times. Over this same period, the rural population increased only slightly, from 445,000 to 553,000, or by 25 %. The percentage of the population living in cities increased significantly, from around 60 % in the early 1950s to 84.2 % in 1970 and 91.7 % in 2005.

**Poverty is a serious problem**

At present, 40 % of Israel children are facing poverty, and 40 % of Israelis believe that it is difficult to live properly on their wages. Both the employment rate and education level are high, but the high cost of living and low incomes have created a degree of poverty. In 2011, the average monthly income of an Israeli family was about $579, compared with $517 in 2010 (NBSC 2012).

## 2.7  Pakistan

### 2.7.1  Overview of Pakistan

Pakistan starts from the frozen Pamirs in the north and extends to the temperate Arabian beaches in the south, stretching over 2000 km along the Indus River. Pakistan neighbors China to the north, Afghanistan to the northwest, Iran to the west, and India to the east. The land area exceeds 800,000 km². The northern part of the country contains several great mountain ranges—the Karakoram, Himalaya, and Hindukush—with some of the highest peaks in the world.

Pakistan's population is 185 million in 2014, making it the world's sixth-most-populous country (NBSC 2015). Punjab Province has the highest population density, and Baluchistan province the lowest. The urban population continues to grow, accounting for over 30 % of the total population in 2014. Pakistan is located at the intersection of Central Asia, South Asia, and Southeast Asia, and this strategic position gives the country great potential for becoming an important center of economic activity. In addition to the existing extensive railway and highway networks, the government is constructing Gwadar Port and highways connecting neighboring areas. Pakistan is a developing country and the government has tried to prioritize the development of both the economy and society (Fig. 2.10).

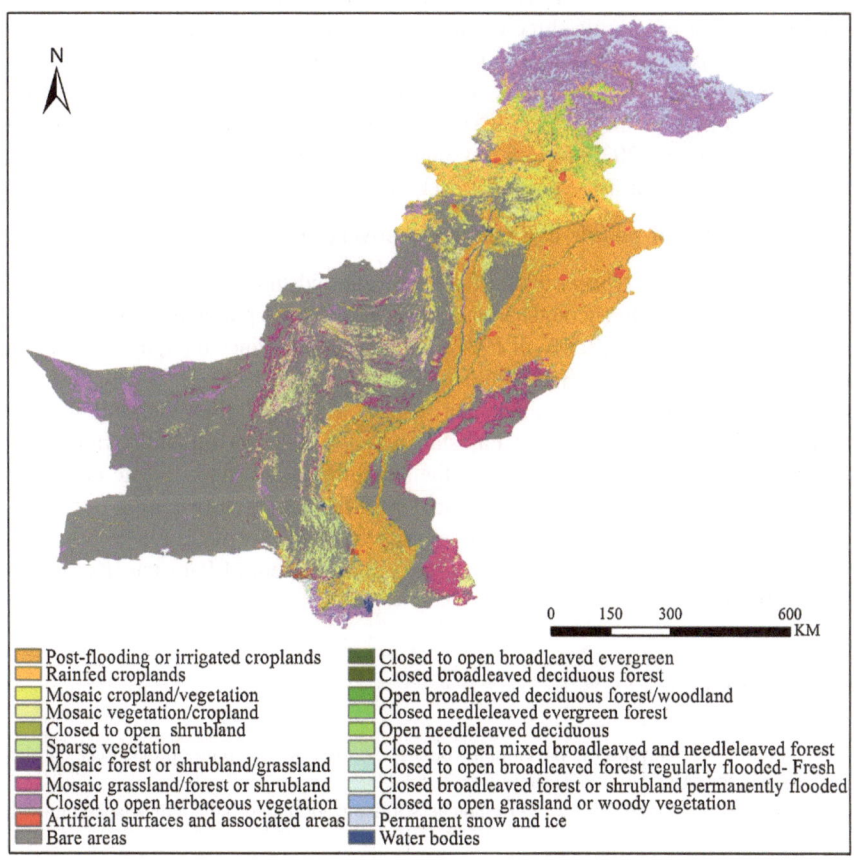

0    150    300    600
KM

| | |
|---|---|
| Post-flooding or irrigated croplands | Closed to open broadleaved evergreen |
| Rainfed croplands | Closed broadleaved deciduous forest |
| Mosaic cropland/vegetation | Open broadleaved deciduous forest/woodland |
| Mosaic vegetation/cropland | Closed needleleaved evergreen forest |
| Closed to open shrubland | Open needleleaved deciduous |
| Sparse vegetation | Closed to open mixed broadleaved and needleleaved forest |
| Mosaic forest or shrubland/grassland | Closed to open broadleaved forest regularly flooded- Fresh |
| Mosaic grassland/forest or shrubland | Closed broadleaved forest or shrubland permanently flooded |
| Closed to open herbaceous vegetation | Closed to open grassland or woody vegetation |
| Artificial surfaces and associated areas | Permanent snow and ice |
| Bare areas | Water bodies |

**Fig. 2.10** Land use map of Pakistan (*Data source* European Space Agency 2009)

### 2.7.2 Urbanization Rate of Pakistan

The urbanization rate has increased slowly in Pakistan, rising from 17 % in 1951 to 36.6 % in 2010. The annual average growth rate between 1991 and 2010 was 3.1 %, compared with the 2.7 % average rate for the South Asian area in the same period. Since 2010, the urbanization rate has increased steadily to reach 38.3 % in 2014 (Rauf et al. 2015).

### 2.7.3 The Characteristics and Problems of Urbanization in Pakistan

#### The high population growth rate

At present, the population growth rate in Pakistan is one of the highest in both Asia and the world. The population has doubled in the last 26 years. At the beginning of Pakistan's independence in 1947, the increase of one million people took about a year, and now it takes only three months. According to the estimate of the United Nations Population Projections and Estimates, Pakistan's population will reach 260 million by 2025. However, unless the government has a clear population development strategy and implements it successfully, the population is more likely to reach 280 million by 2025. If such a strategy were to be fully implemented, then the population of Pakistan may reach a lower figure of 246 million by 2025.

#### A large number of people live below the poverty line

Pakistan's *Economic Observer* (1998–1999) reported that in 1969–1970, the percentage of poor people accounted for 46.5 % of the total population in Pakistan, a figure that decreased to 17.3 % in 1987–1988. However, since then, it has started to increase, accounting for 22.3 % of the total population in 1992–1993 and 29.0 % at present. At least 41 million people live below the poverty line, without sufficient food and clothing. In 1997, the Pakistan human development index ranked the nation 138th out of 174 countries, and the situation is worrying. People in poverty are unable to obtain good education, stable work, effective participation in politics, and social life. If this situation cannot be resolved, then poverty is likely to become one of the greatest threats to the security and development of Pakistan (Komal and Abbas 2015).

# References

Cohen B (2006) Urbanization in developing countries: current trends, future projections, and key challenges for sustainability. Technol Soc 28:63–80

Das S, Ghate C, Robertson PE (2015) Remoteness, urbanization, and India's unbalanced growth. World Dev 66:572–587

Dewan AM, Yamaguchi Y (2009) Land use and land cover change in Greater Dhaka, Bangladesh: using remote sensing to promote sustainable urbanization. Appl Geogr 29:390–401

European Space Agency (2009). http://www.esa.int/ESA

Firman T (2009) The continuity and change in mega-urbanization in Indonesia: a survey of Jakarta-Bandung Region (JBR) development. Habitat Intl 33:327–339

Fahmi FZ, Hudalah D, Rahayu P, Woltjer J (2014) Extended urbanization in small and medium-sized cities: the case of Cirebon, Indonesia. Habitat Intl 42:1–10

Goldblum C, Wong T-C (2000) Growth, crisis and spatial change: a study of haphazard urbanisation in Jakarta, Indonesia. Land Use Policy 17:29–37

Komal R, Abbas F (2015) Linking financial development, economic growth and energy consumption in Pakistan. Renew Sustain Energy Rev 44:211–220

Jang M, Kang C-D (2015) Urban greenway and compact land use development: a multilevel assessment in Seoul, South Korea. Landsc Urb Plann 143:160–172

Jeong SK, Ban YU (2014) The spatial configurations in South Korean apartments built between 1972 and 2000. Habitat Intl 42:90–102

Kim K-H, Pauleit S (2007) Landscape character, biodiversity and land use planning: the case of Kwangju City Region, South Korea. Land Use Policy 24:264–274

Lan Q, Zhang Q (2013) Japanese urban development and its implication for China. City 8:34–37 (in Chinese)

Martinez-Fernandez C, Weyman T, Fol S, Audirac I, Cunningham-Sabot E, Wiechmann T, Yahagi H (2016) Shrinking cities in Australia, Japan, Europe and the USA: from a global process to local policy responses. Progr Plann 105:1–48

Moghadam HS, Helbich M (2013) Spatiotemporal urbanization processes in the megacity of Mumbai, India: a Markov chains-cellular automata urban growth model. Appl Geogr 40:140–149

Murakamia A, Zain AM, Takeuchi K, Tsunekawa A, Yokota S (2005) Trends in urbanization and patterns of land use in the Asian mega cities Jakarta, Bangkok, and Metro Manila. Landsc Urb Plann 70:251–259

NBSC (2012) International statistical yearbook. China Statistics Press, Beijing

NBSC (2015) International statistical yearbook. China Statistics Press, Beijing

Parka S, Jeonb S, Kim S, Choi C (2011) Prediction and comparison of urban growth by land suitability index mapping using GIS and RS in South Korea. Landsc Urb Plann 99:104–114

Pandey B, Seto KC (2015) Urbanization and agricultural land loss in India: comparing satellite estimates with census data. J Environ Manage 148:53–66

Rauf O, Wang S, Yuan P, Tan J (2015) An overview of energy status and development in Pakistan. Renew Sustain Energy Rev 48:892–931

Sharma A, Chandraskekhar S (2014) Growth of the urban shadow, spatial distribution of economic activities, and commuting by workers in rural and urban India. World Dev 61:154–166

Sorensen A (2016) Periurbanization as the institutionalization of place: the case of Japan. Cities 53:134–140

Swerts E, Pumain D, Denis E (2014) The future of India's urbanization. Futures 56:43–52

Wang Y, Chen L, Kubota J (2016a) The relationship between urbanization, energy use and carbon emissions: evidence from a panel of Association of Southeast Asian Nations (ASEAN) countries. J Clean Prod 112:1368–1374

Wang Y, Li L, Kubota J, Han R, Zhu X, Lu G (2016b) Does urbanization lead to more carbon emission? evidence from a panel of BRICS countries. Appl Energy 168:375–380

Wu R, Geng Y, Liu W (2016) Trends of natural resource footprints in the BRIC (Brazil, Russia, India and China) countries. J Clean Prod

# 3    Comparison of Urbanization Processes for Asian Countries

## 3.1    The Temporal Dimension

The different stages of urbanization in the selected Asian countries are summarized in Fig. 3.1 and the urbanization rate trajectories for the same countries are shown in Fig. 3.2. As shown, the characteristics of the various stages of urbanization differ between these countries (Zhang and Seto 2011). In terms of urbanization rate trajectories, South Korea is unique as its urbanization rate increased rapidly within a short period. China, India, Pakistan, and Indonesia can be categorized into one group because their urbanization rates were around 20 % in the 1960s, and although the rates are still not very high compared to developed countries, these four Asian nations have started to undergo very rapid urbanization. Urbanization rates in Russia and Israel have remained stable in recent years, although at different levels.

## 3.2    Sizes and Spatial Structures of Urban Areas

The beginning of Japan's urbanization was characterized by the development of several large cities, with the urban population being highly concentrated with respect to its geographic distribution. This meant that Japan embarked on a path towards a highly compact type of urbanization. After the Meiji Restoration, Japan's industrial development pattern, along with its particular combination of politics and business, coupled with limited land availability, poor resources, and other limiting conditions, made Japan's urban population distribution become highly concentrated, forming a city pattern with "Tokyo metropolitan circle", "Osaka City Circle," and "Nagoya metropolitan circle" as the main body of urban development. Figure 3.3 shows that the metropolitan area (first-class metropolitan area) along the Pacific coast accounts for 72.6 % of Japan's national land area, and includes "capital circle", "central circle", "capital city environs circle", and three secondary metropolitan circles, together accounting for 26.7 % of Japan's population. The core area, including "Tokyo circle", "Nagoya circle", and "Kansai circle" (the three metropolitan areas), which together account for 10.3 % of Japan's national land area, contains 46.7 % of the Japanese population. Of these metropolitan circles, the capital circle has the densest population, with a density of 1317 people per km$^2$, 1.28 times and 1.84 times the densities of the central circle and capital city environs circle, respectively. These urban circles contain most of Japan's population and economic activity within a very small geographic area, and play a decisive role in the country's economic activities.

Regarding the urbanization of the Soviet Union, the Soviet Union Construction Commission defines the following types of city: a city with 500,000 to one million people is a megacity; a city with 200,000–500,000 people is a large city; with

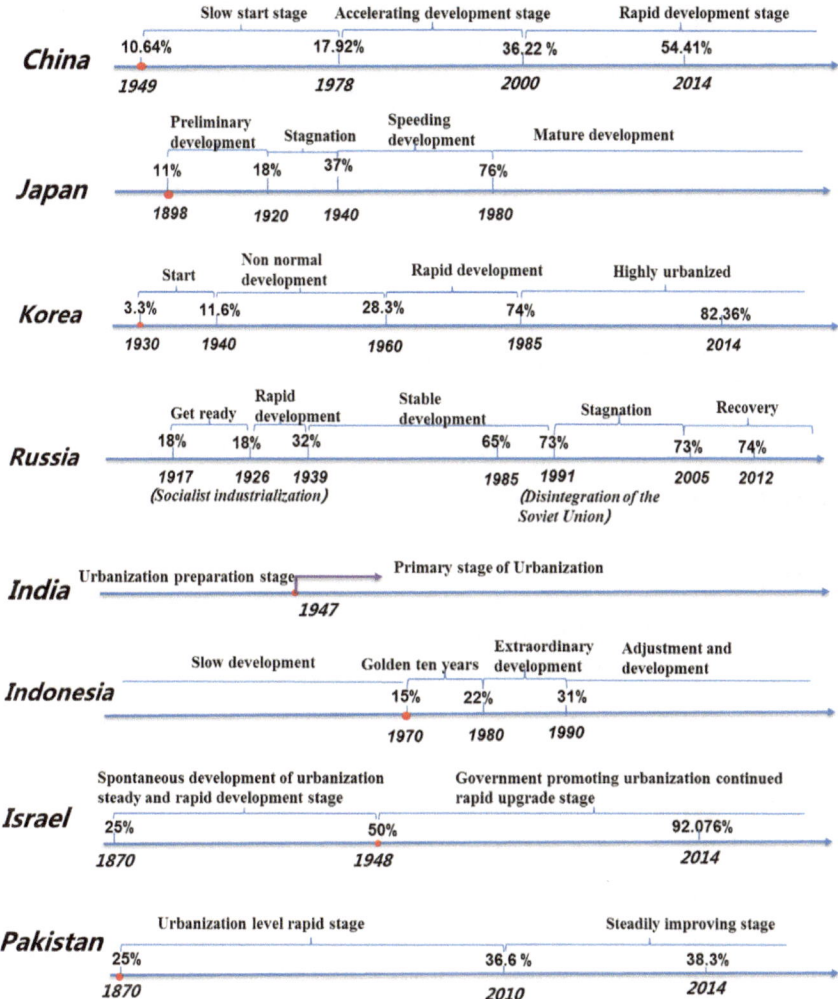

**Fig. 3.1** Stages of urbanization and related values of urbanization rate for eight countries in Asia (*Data source* World Bank)

100,000–200,000 inhabitants is a big city; with 60,000–100,000 people is a medium-sized city; and with 10,000–60,000 inhabitants is a small city. Table 3.1 shows the growth in the number of these types of Soviet city between 1926 and 1986, and Fig. 3.4 shows the rate of growth by city type. In Fig. 3.4, dashed line is for small cities and the growth rate is labeled in the right axis. The growth rates for the other categories of cities are labeled in the left axis.

Comparing the number and growth rate of the different types of Soviet city, it is clear that the proportion of small cities is high and that the number of large cities has grown. The urbanization policy of the Soviet Union can be termed the

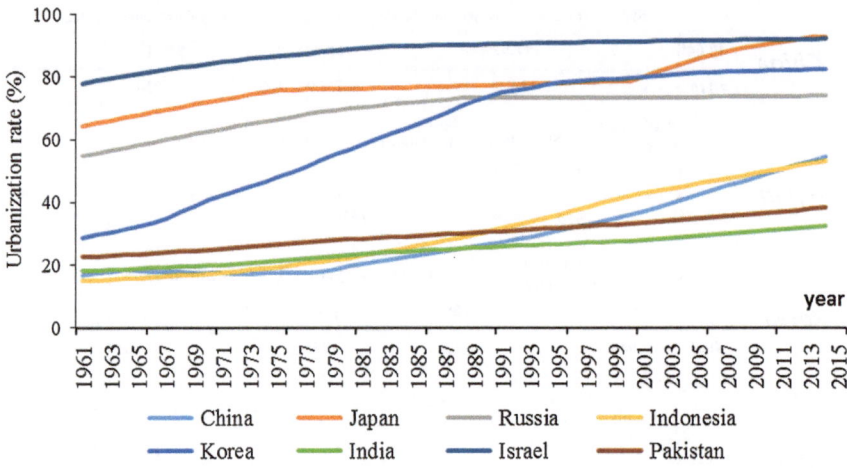

**Fig. 3.2** Changes in urbanization rate for eight countries in Asia (*Data source* World Bank)

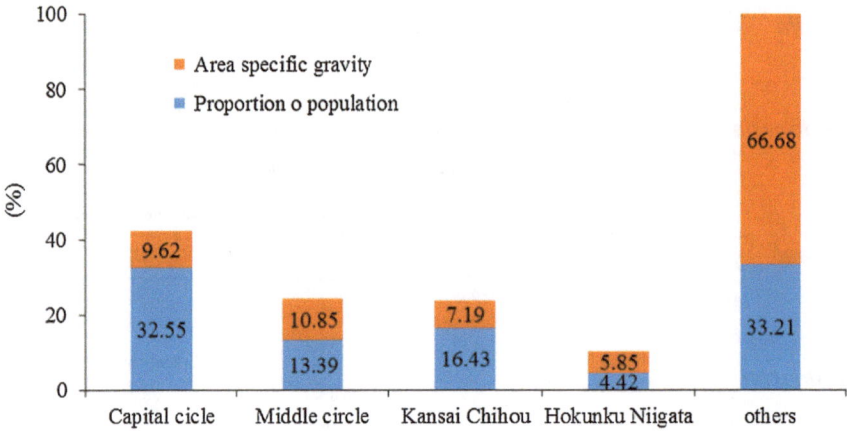

**Fig. 3.3** Japan's major urban agglomeration areas and percentage of the population contained therein (*Data source* Japan National Bureau of Statistics)

"scattered concentration" strategy, a strategy that was implemented to avoid the excessive concentration of population and industry. Although the control of extensive urban growth did not achieve satisfactory results, the policy of encouraging the development of small and medium-sized cities (especially new cities in the eastern part of the Soviet Union) and the establishment of a large number of towns did result in a large increase in the number of smaller cities and towns.

Israel's major cities will continue to play a dominant role in the distribution of population and economic activity of the nation in the future. Although the populations of Jerusalem, Tel Aviv, and Haifa have been reduced over time from 75 to

**Table 3.1** The growth in the number of Soviet cities

| Year | 1926 | 1939 | 1959 | 1979 | 1985 | 1986 |
|---|---|---|---|---|---|---|
| Total number | 709 | 1194 | 1679 | 1941 | 2085 | 2190 |
| Small city | 618 | 1007 | 1375 | 1519 | 1592 | 1654 |
| Medium city | 60 | 98 | 156 | 200 | 221 | 239 |
| Big city | 78 | 78 | 123 | 188 | 226 | 243 |
| Large city | 1 | 9 | 22 | 25 | 26 | 32 |
| Megacity | 2 | 2 | 3 | 9 | 20 | 22 |

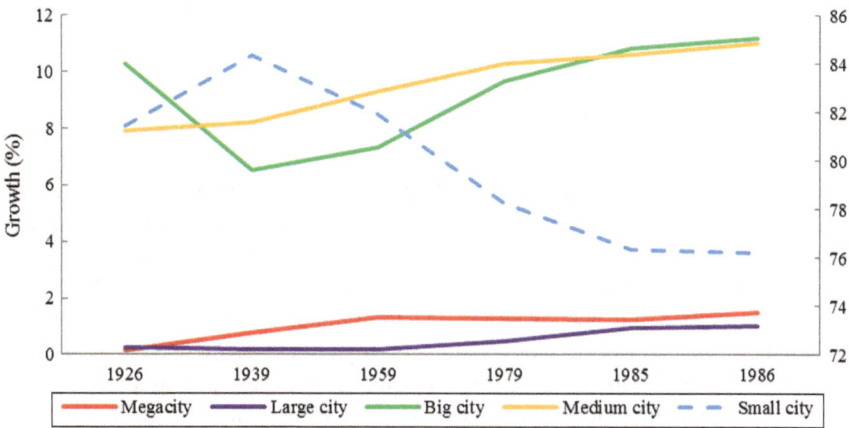

**Fig. 3.4** The rate of growth of the former Soviet Union by city type (*Data source* C. Brook, World Population)

45 % of the total population of the Israel, these cities have retained their economic power, human resources, technology, information, capital, transportation, and other advantages that they had previously established. The continued development and function of the big cities not only boosted economic development, but also stimulated the development of the rural economy. The big cities, including Tel Aviv and Haifa, became the "growth pole" of the nation's economy and social development. These larger cities were not only the national industrial development center, transportation center, commodity circulation center, and science and technology center of the country, but also the financial center, talent-gathering center, and regional integrated service center. The summary of urban scale structure for these selected Asian countries is presented in Table 3.2.

**Table 3.2** Summary of urban scale structure for seven Asian countries

| Country | Urban scale structure |
|---------|----------------------|
| Japan | Coordinated development of large, medium, and small cities |
| Korea | Unbalanced spatial development, the status of large cities was prominent, and the population of the capital city region was over-concentrated |
| Russia | Small cities were numerous and large cities grew in both size and number |
| India | The structure of the inverted triangle, whereby cities and metropolitan areas developed rapidly, medium-sized cities developed slowly or stagnated, and small cities were in recession |
| Indonesia | A focus on the development of medium-sized and small cities |
| Israel | The establishment of new towns/cities was an important urbanization strategy |
| Pakistan | The level of urbanization increased slowly |

## 3.3 The Layout of Urban Space

Japan's overall level of urbanization is high. The nation's limited areas of suitable land, complex terrain, and resource constraints mean that the population and economic activity are highly concentrated, particularly along the eastern coast of the island of Honshu. Japan's cities are concentrated mainly in the Kanto Plain near Tokyo, the Nobi Plain near Nagoya, and the Guinea Plain near Osaka, forming three major urban clusters of Tokyo circle, Nagoya circle, and Kobe circle. In 2001, the economies of the three major urban clusters collectively accounted for 48.6 % of the gross population and created 66.2 % of the GNP. The area covered by these three urban clusters is only 10.4 % of the whole country, but they have become the main driver of Japan's economy. The industrial structures of the city groups are different from those of other cities, and their functional characteristics are distinctive. The capital circle with Tokyo as the center is not only a large market and a very important industrial belt, but is also a world-leading economic, financial, and trade center. The Keihanshin area with Osaka, Kobe, and Kyoto as the center was the largest market in the Edo period (1603–1867), and now forms a large industrial zone based on the production of consumer goods. Tokyo Metropolitan Area, with Nagoya as the center, gradually developed into a heavy industry zone, and is the largest heavy industry and chemical manufacturing/processing area in Japan.

The level of urbanization in South Korea has increased rapidly, being characterized by the development of a few large cities with subsequent diffusion. South Korea's large cities are concentrated in the transportation hub (Daejeon and Daegu), coastal ports (Busan, Incheon), the metropolitan area (An Shan, Cheng Nan, Gao Yang, and others), and the industrial center area (Ma Shan, Chang Yuan, and others). South Korea faced a shortage of both resources and capital at the beginning of its economic modernization; therefore, the nation had to focus on those areas with more advantageous locations and with conditions favoring development, which led to the unbalanced distribution of urbanization and economic activity.

South Korea's political regionalism has meant that the Gyeongsang region has been more successful, compared with other regions, in obtaining government funding and infrastructure. This has added to the unbalanced development of urbanization in South Korea. At present, about 25 % of South Koreans live in Seoul, and the combined population of the five cities of Seoul, Busan, Daegu, Inchon, and Kwangju accounts for more than half of the population of the country.

Russia's urban areas are concentrated mainly in the western part of the country, although are represented in a more restricted way in the central and eastern areas. Russia's urbanization development started in the west, and after the 1917 revolution, the state implemented strategies to develop urban areas further eastward. However, because of the collapse of the Soviet Union, the density of cities in the eastern part of the country is still low. In the western part of the territory formerly covered by the Soviet Union, there are 10–17 cities per 10,000 km$^2$. In the eastern part, there are only 0.6 cities per 10,000 km$^2$, and in the northern part of Siberia and the far eastern region there is only 1 city every 0.2 to 0.5 million km$^2$. The 1992 and 2013 night-time light data from the Operational Linescan System (OLS) of the Defense Meteorological Satellite Program (DMSP) for Russia confirm these patterns, indicating that Russian towns and cities are concentrated mainly in the western parts of the country, and gradually spread to the east over the 20-year period, which illustrated in Figs. 3.5 and 3.6 (Zhang and Seto 2011).

In India, a small number of large cities and megalopoli are located in particular areas, apparently isolated and lacking contact with the smaller cities of the country (Taubenböck et al. 2009). In Indonesia, the urban areas are concentrated mainly in the corridor region of Java–Sumatra–Jakarta–Barry–Dong. In Israel, urban growth is concentrated mainly in the northern part of the country and in coastal areas, and from 1992 to 2012, the level of urbanization increased quite substantially, the number of towns in the northern part of the country increased, and several cities emerged in the southern part. In Pakistan, urbanization occurred at first mainly in the southeastern part of the country, spreading to surrounding regions over time. The summary of the layout of urban space, in terms of geographic distribution and spatial pattern, is presented in Table 3.3 (Taubenböck et al. 2012).

## 3.4 The Coordination of Urbanization and Industrial/Economic Development

Increasing levels of urbanization require a corresponding increase in tertiary industries, in particular information and business services (Hsieh 2014).

We selected the industrial added value and the proportion of GDP in the same year to measure the industrialization rate (I), and the ratio of I to the urbanization rate (or level) (U) was used as an indicator of complementarity between industrialization and urbanization. Japan's urbanization occurred simultaneously with industrialization (Fig. 3.7a). The industrialization/urbanization (IU) ratio measured

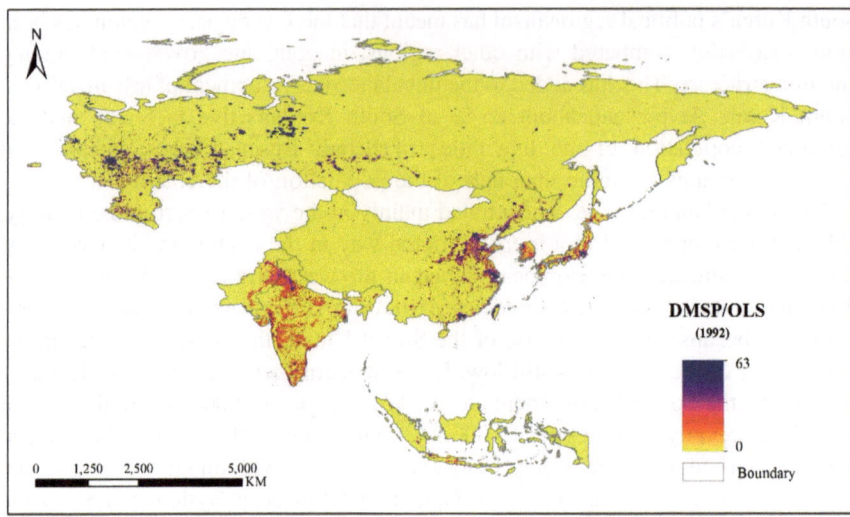

**Fig. 3.5** Night-time lights data for selected Asian countries from DMSP/OLS in 1992

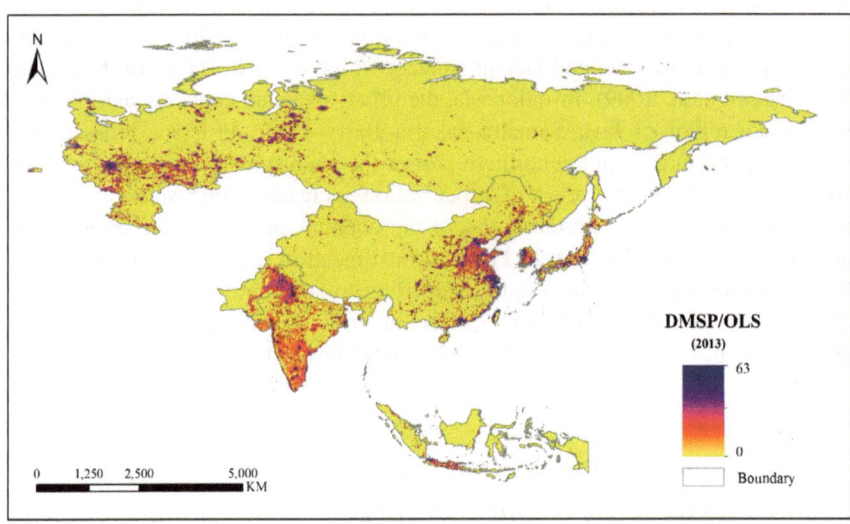

**Fig. 3.6** Night-time lights data for selected Asian countries from DMSP/OLS in 2013

**Table 3.3** Summary of the layout (geographic distribution and pattern) of urban space

| Country | Layout of urban space |
| --- | --- |
| Japan | A high degree of urbanization along the eastern coast of Honshu, and urbanization in different areas developed in a coordinated way |
| Korea | Concentration in a few large cities, then diffusing outwards, particularly in Seoul and Busan |
| Russia | Most cities are located in the western part of the country, although with some decentralization towards the east |
| India | A small number of large cities and megalopoli are located in particular parts of India, but are relatively isolated and lack contact with small cities |
| Indonesia | Urban areas are concentrated in the corridor region of Java, Sumatra–Jakarta–Barry–Dong |
| Israel | Urban development was concentrated in the coastal and central parts of the country, with cities and towns increasing in number and enlarging in the northern part of the country |
| Pakistan | Urbanization took place at first mainly in the southeastern part of the country, subsequently spreading to the surrounding regions |

about 0.5 during the 1970s and 1980s, indicating the simultaneous development of urbanization and industrialization and their complementary nature. Japan's IU ratio has declined consistently over time, from 0.6 in 1970 to 0.3 in 2012. The trend of urbanization level and industrialization level is similar basically. Although the level of urbanization and economic development in Japan were both high (Fig. 3.7a), the level of urbanization was slightly lower than the level of economic development, indicating that Japan was built upon a relatively affluent mode of urbanization. The level of urbanization and the level of economic development were comparatively coordinated.

South Korea underwent urbanization that was coordinated with industrialization. The level of urbanization was high, and the average IU ratio was more than 0.5, although it varied between 0.65 and 0.46, gradually decreasing over time. The pattern (Fig. 3.7b) shows that coordinated development of urbanization and industrialization occurred, and that the process of urbanization was very rapid. South Korea, over a period of 30 years or so, experienced an unprecedented rapid pace of urbanization and became a highly urbanized country, but the pace of industrialization was relatively slow. Russia also underwent coordinated urbanization (Fig. 3.7c). Over time, the IU ratio declined to about 0.5, indicating Russia's coordinated development of urbanization and industrialization. The level of urbanization was the same as the level of economic development.

India underwent lagged urbanization (Fig. 3.7d). The IU ratio is very high, measuring between 0.9 and 1.1 since 1969, indicating that urbanization lagged behind industrialization. The speed of India's urbanization was slow and the main reason for this was that the government wanted to avoid the antagonism between urban and rural areas and "city disease", including problems such as serious traffic congestion and high levels of air pollution from factories.

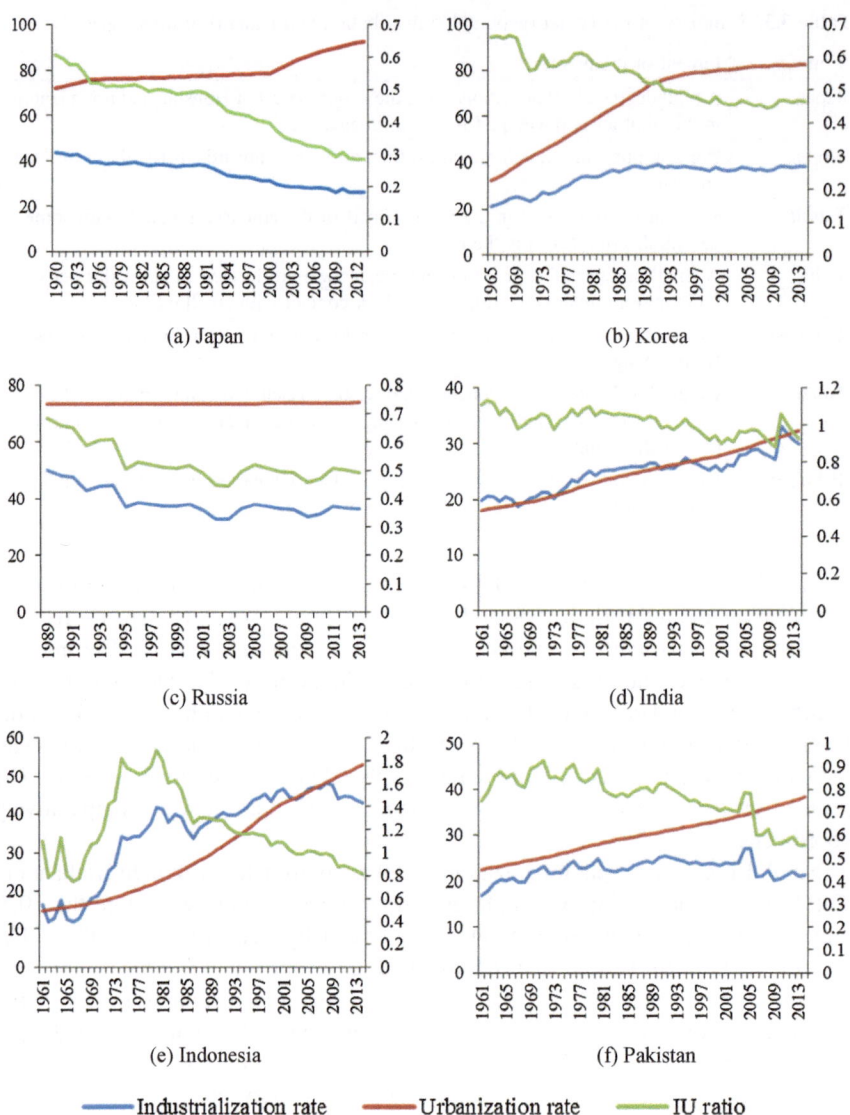

**Fig. 3.7** Coordination between urbanization and industrialization (*Data source* World Bank)

Indonesia also underwent lagged urbanization (Fig. 3.7e). The IU ratio was generally high, reflecting that urbanization lagged behind industrialization and that the urbanization level was low. However, in more recent years, the level of urbanization has gradually increased and the IU ratio has gradually declined. The levels of urbanization and economic development of Indonesia were both low but the level of urbanization was lower than the level of economic development. The level of urbanization and the level of economic development were not coordinated, meaning that Indonesia experienced lagged urbanization.

Pakistan has been transitioning from lagged urbanization to coordinated urbanization (Fig. 3.7e). Before 2006, the IU ratio was generally high (0.8–0.9), reflecting that urbanization lagged behind industrialization and that industrialization developed essentially on its own, unrelated to urbanization. The IU ratio has dropped to around 0.5 since 2006. Pakistan's industrialization rate was low, indicating that the levels of industrialization and economic development of Pakistan were very low. The turbulent domestic environment led to the creation of a large number of poor people. The level of urbanization has been increasing year by year, but has not been coordinated with the level of industrialization. Rather, urbanization occurred on the basis of relative poverty and economic backwardness. The industrialization process was slow, but the country is now in the process of changing from lagged urbanization to coordinated urbanization. The summary of the coordination of development for these selected Asian countries is presented in Table 3.4.

## 3.5 Dynamic Mechanisms

Japan's urbanization can be examined in terms of three major dynamic mechanisms.

**Industrial layout policies and measures** Although Japan is a market economy country, the government plays an important role in industrial development and urban layout. Since the Meiji era, the central government has established national industrial enterprises as a "navigation factory", and subsequently transferred these to the private sector, but the government had been providing subsidies, protection

**Table 3.4** Summary of the coordination of development

| Country | Coordination of development |
|---------|----------------------------|
| Japan | Coordinated urbanization: the coordinated development of industrialization and urbanization |
| Korea | Coordinated urbanization: the fluctuation in the ratio of industrialization to urbanization is small |
| Russia | Lagged urbanization: industry relies too much on energy and consumption industries, the comparative levels of urbanization and industrial development are not reasonable |
| India | Lagged urbanization: the level of urbanization lags behind the levels of industrialization and economic development |
| Indonesia | Advanced urbanization: the level of urbanization depends on secondary and tertiary industries |
| Israel | Coordinated development of urbanization |
| Pakistan | The transition from lagged urbanization to coordinated urbanization |

measures, and tax preferences. After WWII, the government provided significant funds for the reconstruction of industry. To ensure prioritized growth in economic sectors, it was particularly important in the 1950s and 1960s that both central and local government provided a wide variety of financial support and assistance regarding industrial construction land, the preparation of industrial zones, industrial water, the construction of transport facilities, and technical assistance of various types. The government put its energy into developing the economy based on export industries, and this export-oriented economic strategy meant that industries became highly concentrated in the coastal cities.

**Promotion of the coordinated development of urban and rural areas by legal means** Japanese decision-makers noticed agricultural and rural development issues in the late stages of urbanization, and enacted a number of laws to promote rural development. For instance, with respect to supporting the healthy development of rural mountainous areas and regions with low numbers of people, the laws included the Special Active Measures Law in Depopulated Areas, the Peninsula Promotion Act, the Mountain Revitalization Act, the Special Measures Law in The Snow Area, and the Islands Promotion Act. The laws to ensure full employment of the labor force and the establishment of industrial and commercial enterprises in rural areas included the Promotion Law of The Introduction of Industry in Rural Areas, the New Business Innovation Promotion Law, and the Law of Promoting Regional Construction in Small Cities and Re-layout of the Industrial Business Construction. Under the guidance of policy, rural areas underwent great changes, characterized no longer by only farmers living in an area but by specialized households and farmers and non-farmers all mixed in the same community. Agriculture was no longer the dominant industry in the countryside, and the percentage of workers in tertiary industries in Japan's countryside was as high as 42 % in 1980, which was much higher than the percentage working in agriculture (24 %). Small local cities with populations of 10,000–100,000 developed rapidly.

**An increase of investment in rural areas and promotion of the integration of urban and rural areas** The Japanese government at different levels has attached great importance to investment in rural areas. Japan has found many ways and channels for investing in rural areas. The central government allocates funds or loans for construction projects. In addition to these financial allocations, local governments can issue local bonds for the construction of public facilities. The government has invested heavily in rural infrastructure construction: 1.084 trillion yen in 1998 rising to 1.091 trillion yen in 1999. The improvement of rural infrastructure has strengthened the connection between urban and rural areas, and has made the integration of urban and rural areas possible. Rural development has also provided support for the development of urban industry and population migration.

Industrialization has been the main driving force of South Korea's urbanization. In the past 50 years, South Korea has seized the historical opportunity of the industrial transition of developed countries by forming its own industrial chain based on applying advanced technology. This involved participating in

high-added-value activities in the international manufacturing chain, thereby real-
izing the shift to the developed industrialized countries. The enhanced competitiveness of manufacturing and service industries has won a broad international
market for South Korea, has created ample employment opportunities for the South
Korean people, and has increased the level of urbanization in the nation. The
modern development of manufacturing industries has placed more complex and
diversified demands on the production service industry, leading to the development
of service industries such as finance, insurance, business information, communi-
cation, logistics, management consulting, advertising, and engineering, with an
emphasis on meticulous division and specialization. Rising worker incomes have
also stimulated consumption and the development of social security services, and
have improved the quality of urban life in South Korea.

Before the dissolution of the former Soviet Union, the urbanization of Russia
was dominated by the plans and actions of the government. After the collapse of the
Soviet Union, the Russian Federation became independent and moved towards
capitalism, and urbanization was dominated by the market.

The Israeli government has promoted the development of cities by formulating
appropriate policies for particular areas. According to statistics, in 1990 Israel's
urbanization rate was 90.3 %, whereas the urbanization rate of Bedouin was about
59 %; in 2005, the national rate of urbanization was 91.7 %, and the urbanization
rate of Bedouin was 76 %. These figures reveal that the urbanization gap has been
gradually narrowing in Israel, with better integration between urban and rural areas.
The summary of dynamic mechanisms of urbanization for these selected Asian
countries is presented in Table 3.5.

## 3.6 Industrial Sector Employment

The contribution to GDP made by each of the main industrial sectors (primary
industry or agriculture, secondary industry or manufacturing, and tertiary industry
or service industry) accounts for the total GDP. Similarly, the percentage of
employees working in each of the main industrial sectors accounts for the total
employment. Changing patterns of industrial contribution to GDP (Fig. 3.8) and
industrial sector employment (Fig. 3.9) over time indicate structural and other
changes in national economies.

Japan gives priority to secondary and tertiary industries, and these industries
play a dominant role in economic development. Since 1960, the contribution of
service industries to the national GDP has increased from 50 to 75 %, with a
corresponding decrease mainly in the contribution by secondary industries
(Fig. 3.9a). Between 1950 and 1958, the number of employees in primary industries was more than that in secondary and tertiary industries, which means that
Japan's labor force was working predominantly in agricultural activities during this
period (Fig. 3.10a). From 1960 to 2000, the number of employees in service
industries increased and became much higher than the numbers employed in

**Table 3.5** Summary of dynamic mechanisms of urbanization

| Country | Dynamic mechanism |
|---------|-------------------|
| China | Transformation of industrial structure and institutional change together promote the process of urbanization |
| Japan | Government-sponsored |
| Korea | Rapid development under the control of both the market and the government. The economy affects urbanization first, infrastructure affects the urban system, and government policies affect both economic development and urbanization |
| Russia | Dominated by government plans and actions before the dissolution of the Soviet Union and by the market after the creation of Russia as an independent federation |
| India | Affected almost entirely by the market, the government's impact is very limited |
| Indonesia | Dominated by the market, the government has little effect |
| Israel | Government-sponsored |
| Pakistan | Government-sponsored |

primary or secondary industries, representing a movement of labor from the primary sector into both the secondary and tertiary sectors, but mainly into the tertiary sector.

Before the transition of Russia from a member of the Soviet Union to an independent federation, development in agriculture, light industry, and heavy industry was unbalanced. Excessive development was made in heavy industry, and agriculture and light industry both fell behind. According to official information released by the Soviet Union, if the index was 1 in 1913, the industrial (secondary industry) output value was 7.7 in 1940, 92 in 1970, 163 in 1980, and 195 in 1985. The equivalent figures for heavy industry are 1, 13, 214, 391, and 468, respectively; for light industry are 1, 4.6, 30, 50, and 61, respectively; and for agriculture are 1, 1.4, 3.1, 3.4, and 3.8, respectively. Concerning employment structure, according to the statistics of the British economist Clark, in 1913, the percentage of the total labor force accounted for by each of the three productive sectors in Tsarist Russia before the October 1917 Revolution was: primary industries 75 %, secondary industries 9 %, and tertiary industries 16 %. By 1950, the equivalent percentages for the Soviet Union were 48, 27, and 25 %, respectively. Heavy industry accounted for 72.5 % of the secondary industry sector. From 1940 to 1985, industry developed rapidly and the proportion of heavy industry became higher. To obtain more capital to support the rapid development of heavy industry, the Soviet government implemented a policy of lowering the price of agricultural products and light industry products for a long time and relying on the sacrifice of agriculture and light industry in exchange for the prioritized development of heavy industry. After Russia's transition to an independent federation, according to the figures, the development of the three major industries became more balanced. The summary of employment by industrial sector for these selected Asian countries is presented in Table 3.6.

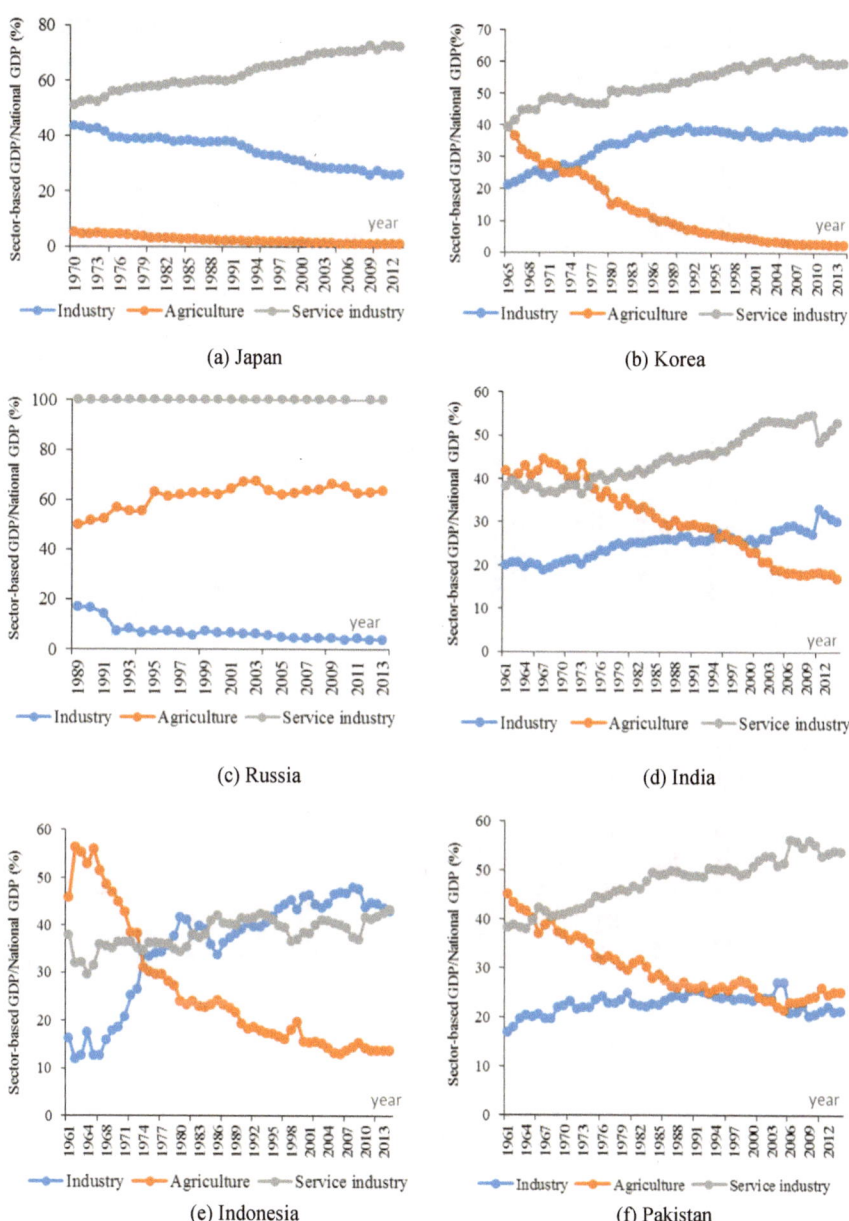

**Fig. 3.8** Percentage contributed to national GDP by each of the three main industrial sectors (*Data source* World Bank)

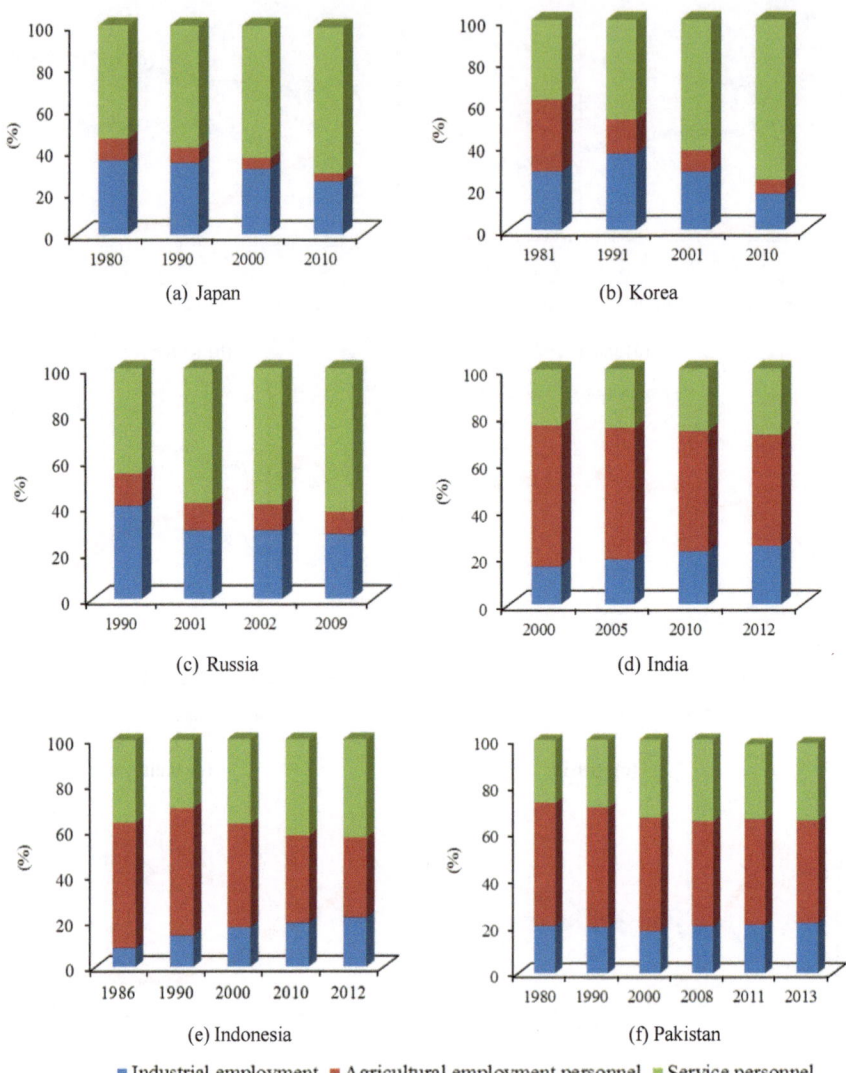

**Fig. 3.9** Employment percentage by industrial sector (*Data source* World Bank)

## 3.7 Rural–Urban Relationships

The capital of Japan is Tokyo, one of the largest metropolitan areas in the world and a city with a very high population density. There are about 8.6 million people living in the Tokyo Metropolitan Area. In the Tokyo special area and surrounds, numerous Fukutoshin have been formed. If each of Tokyo's 23 districts is considered as a city, then the population of 12 of these Japanese cities exceeds one

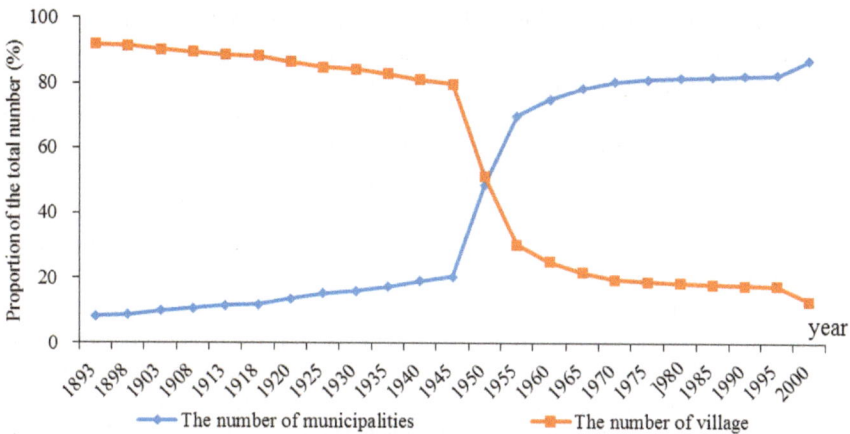

**Fig. 3.10** Changes in the number of Japanese rural municipalities (*Data source* Historical study of Japan's industrialization, urbanization, and the evolution of farmland systems)

**Table 3.6** Summary of employment by industrial sector

| Country | Industrial sector employment |
|---------|------------------------------|
| Japan | The secondary and tertiary sectors dominate |
| Korea | The tertiary sector dominates, the primary sector has declined |
| Russia | The relative development of agriculture, light industry, and heavy industry proportion was unbalanced, with heavy industry being excessively developed |
| India | Pay attention to industrial development, but do not ignore agricultural development |
| Indonesia | The secondary and tertiary sectors dominate, tertiary industries make the biggest contribution to economic activity |
| Israel | Secondary industries and the military industry are the pillar industries, tertiary industries are also strong |
| Pakistan | Tertiary sector industries dominate, primary and secondary industries are attaining a more balanced development |

million. Currently more than 90 % of Japanese live in big cities, medium-sized cities, or towns. Only a relatively small number of people live in rural areas, fishing villages, and mountain villages. The high degree of urbanization appears to guarantee high living standards for residents, and also boosts domestic demand and promotes the development of various types of service industries.

However, the over-urbanization has brought a series of problems to Japan. For example, Japan's urban development is too concentrated. In addition, only a few cities in Tokyo and its surrounding area are still prosperous, and large cities such as Osaka and Kobe are declining; the situation of small towns and cities is more serious. This will lead to more Japanese continuing to move to the very largest cities, especially the capital circle, further reducing populations in other areas (Fig. 3.10). In addition, the Japanese government built a large number of public facilities to promote urbanization in the 1970s and 1980s. Almost all rural areas in Japan had a high-level highway and in many small and medium-sized cities and towns a civic museum, gymnasium, or national resort and other buildings can often be seen. However, these facilities were seldom used, and now the maintenance of these public facilities has placed constraints on local government finances. Many Japanese scholars believe that the government cannot guide urbanization only by building roads and other infrastructure, but needs to find the right industrial direction that suits particular regions and localities. Only when appropriate industrialization is realized and adequate employment opportunities are ensured can satisfactory urbanization be guaranteed. At present, Japan is considering ways to embark on "regional revitalization"; some regions and towns are focusing on the development of characteristic tourism industry activities, some are developing folk arts and crafts, whereas others are building development zones to attract investment. The purpose is to attract younger people back from the cities to re-stimulate the vitality of smaller towns. Given the nation's aging population, some areas are turning to the retired and appealing to them to sell their small houses in the big cities and starting a second life in smaller towns with more attractive natural environments.

To support rapid industrialization, the Japanese government formulated a large number of policies to promote the development of agriculture, first by consolidating municipal amalgamation, and second by attaching importance to the promotion of agricultural operations and agricultural education. Japan's industrialization and urbanization developed in large part through being supported by surpluses in the agricultural sector. On the one hand, agriculture brought a large amount of surplus labor and surplus agricultural products to benefit the development of industrialization and urbanization; on the other hand, arable land exploitation and readjustment and municipal amalgamation provided a territorial resource for industrialization and urbanization. We can conclude that the number of municipalities has been increasing gradually and the number of rural areas has been decreasing. Urban industrial agglomeration areas were gradually being transferred to rural areas, which promoted agricultural development to some extent, and also relieved the pressure of population agglomeration in the large urban areas.

Russia sacrificed the development of its rural areas and instead pursued rapid urbanization and industrialization in the Soviet period. Many of rural inhabitants were attracted to employment in the high-speed industrialization that occurred. The Soviet Union emphasized a reduction in urban–rural disparities, but there were still a large number of rural people who moved into cities given that Soviet agriculture was still in a relatively backward state as a result of the extensive urban and industrial growth mode from the 1920s to the 1950s. According to the 1926 and 1959 censuses, the total number of people who in that period moved from the countryside to a city or those rural inhabitants who became urban residents because of administrative (territorial) changes was 56.1 million, accounting for 80.8 % of the overall urban population. The employment rate of the urban population of the Soviet Union in 1989 was 66 %, and the percentage in agricultural employment was 20 %. The ratio between the two was far from normal. The urbanization of the

**Table 3.7** Summary of rural–urban relationships

| Country | Rural–urban relationship |
|---------|--------------------------|
| Japan | Rural–urban coordinated development |
| Korea | Rural–urban coordinated development |
| Russia | Industrialization and urbanization were pursued at the expense of agricultural (rural) development |
| India | Pay attention to urban and industrial development, but at the same time do not sacrifice the countryside and agriculture |
| Indonesia | Neglect agricultural development and focus on the development of tertiary industry, creating a widening urban–rural gap |
| Israel | Rural–urban coordinated development |
| Pakistan | Pay attention to the development of small towns |

Soviet Union was dominated by the state, and a large number of agricultural migrants had become urban citizens, but because lifestyle and ideas differed between urban and rural areas, these migrants were marginalized. In the few years after the collapse of the Soviet Union, there was an abnormal phenomenon whereby urban residents migrated to rural areas; however, Russia then resumed the urbanization process with people continuing to move to urban areas. The summary of rural-urban relationships for these selected Asian countries is presented in Table 3.7.

## 3.8 Levels of Economic Activity

Japan's final consumption expenditure per capita is higher than that of the other selected countries, with Israel and Korea having medium levels, and China, Pakistan, India, Indonesia, and Russia having relatively low levels (Fig. 3.11). The final consumption expenditure is the market value of all goods and services purchased by residents (including goods such as cars, washing machines, and computers). The statistic does not include the expenditure of house purchase, but does include the estimated rent of owner-occupied housing and the costs of obtaining permits and registrations from the government (Cobbinah et al. 2015). Consumer spending here also includes expenditure made to non-profit organizations for the service of residents.

The 2005 per capita final consumption expenditure (Fig. 3.12) and per capita GDP (Fig. 3.11) show broadly similar patterns. Japanese per capita GDP is relatively high, with that for Israel and South Korea being medium, and low for the other countries. The per capita GDP is an effective reflection of the macroeconomic situation of a country or region. The per capita GDP is the ratio of the GDP in the accounting period (usually a year) to the permanent resident population of the country (or the household registration population). It is a standard for measuring and comparing the living standards of people in different countries.

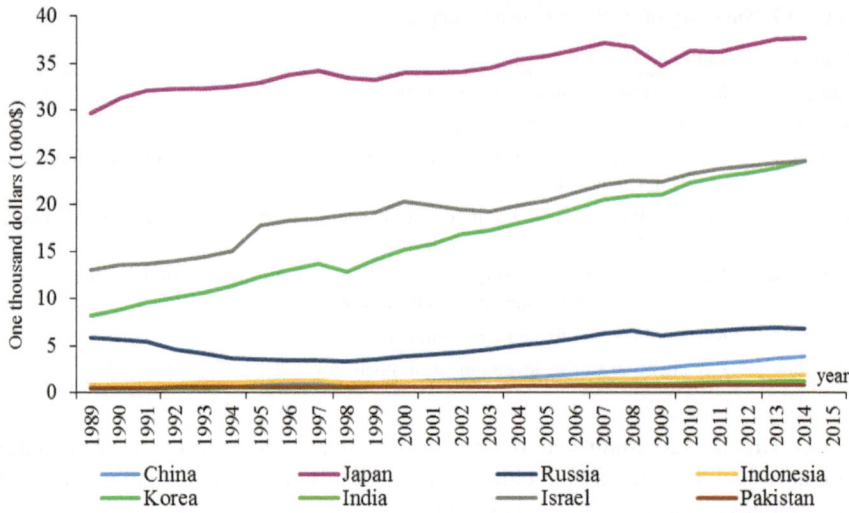

**Fig. 3.11** Changes in per capita GDP for selected Asian countries (*Data source* World Bank)

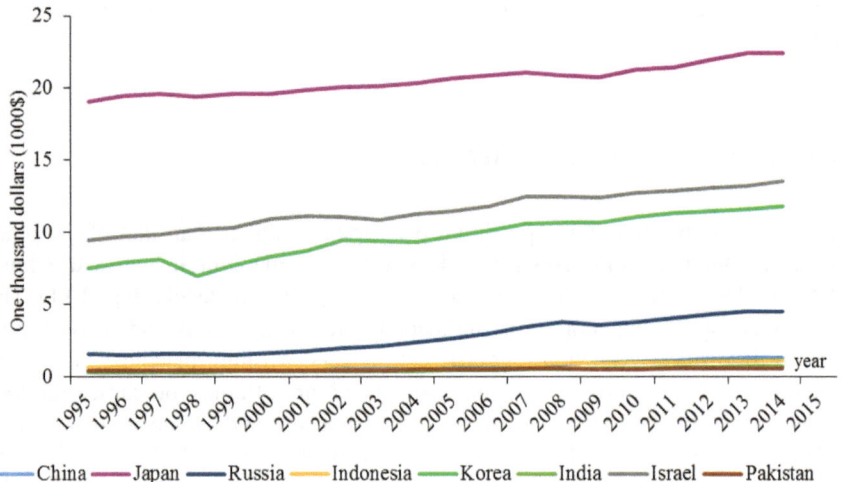

**Fig. 3.12** Changes in residents' final consumption expenditure per capita for selected Asian countries (*Data source* World Bank)

# References

Cobbinah PB, Erdiaw-Kwasie MO, Amoateng P (2015) Rethinking sustainable development within the framework of poverty and urbanisation in developing countries. Environ Dev 13:18–32

Hsieh S-C (2014) Analyzing urbanization data using rural-urban interaction model and logistic growth model. Comput Environ Urban Syst 45:89–100

Taubenböck H, Wegmannb M, Roth A, Mehl H, Dech S (2009) Urbanization in India-spatiotemporal analysis using remote sensing data. Comput Environ Urb Syst 33:179–188

Taubenböck H, Esch T, Felbier A, Wiesner M, Roth A, Dech S (2012) Monitoring urbanization in mega cities from space. Remote Sens Environ 117:162–176

Zhang Q, Seto KC (2011) Mapping urbanization dynamics at regional and global scales using multi-temporal DMSP/OLS nighttime light data. Remote Sens Environ 115:2320–2329

# 4  International Urbanization Processes and Patterns: Implications for China

## 4.1  Typology of Urbanization for the Selected Asian Countries

In different historical stages, under the effect of different international environments, capital conditions, resource conditions, and social systems, each country has stepped out on its own road of industrialization and urbanization.

On the basis of the relationship between urbanization rate and per capita GNI data, the seven selected Asian countries and China can be divided into three types, as shown in Fig. 4.1. The three types are: I—Rapid development country; II—Fast-pursuit country; and III—Medium-income country.

### 4.1.1  Rapid Development Countries: Achievement of Synchronous Increase in the Level and Quality of Urbanization in the Long Term

Japan is one of the world's countries (in addition to those in the UK, North America, and Western Europe) that developed first, and is the most developed country in the Asian region. Beginning with the industrial revolution, urbanization was essentially completed by 1940 and achieved a synchronous increase in the level and quality of urbanization over the long term. The urbanization process in Japan has particular characteristics, as outlined below.

First, driven by the industrial revolution, a positive interaction between industrialization and urbanization was achieved. The industrial revolution established a new industrial economic system, which proved to be an inexhaustible motivating force for promoting the development of urbanization. Urbanization also provided a further basis for industrialization and the market, and improved productive

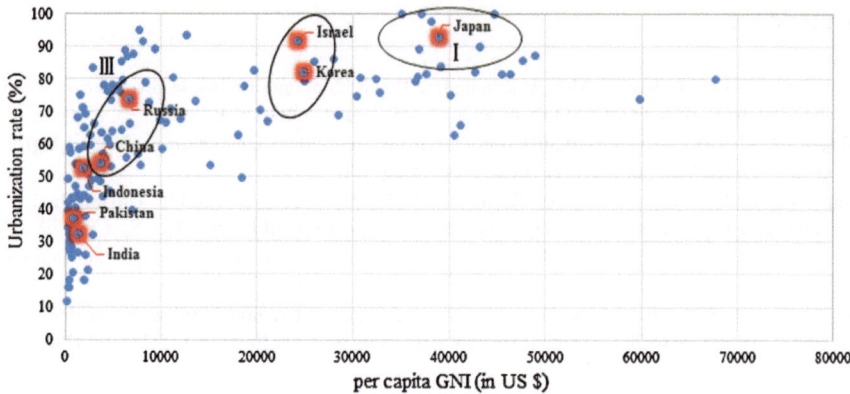

**Fig. 4.1** Three types of country according to urbanization rate and per capita GNI (*Data source* World Bank 2014)

efficiency. The countries of the world that were first to develop gained comprehensive knowledge in technologies and systems, and industrialization and urbanization had generally been integrated for a period of more than 100 years.

Second, the old colonial system and the international distribution of power prevailing at the time provided a particular temporal background to urbanization. On the one hand, the first countries to develop had accumulated a large amount of capital for urbanization through colonial plunder, foreign trade, and wars. On the other hand, in the unequal international economic order and political structure, resources and environmental constraints do not apply to the development of urbanization, and therefore a mode of "high investment, high consumption, and high pollution" and a road of "pollution first, treatment later" was formed. A huge environmental cost has been paid for this approach.

Third, with the gradual improvement of the capitalist system, the government and the market achieved an organic combination in the development of urbanization. In the early stage of urbanization, the free market mechanism plays an important role, and urban development is driven mainly by the private sector rather than by the government. As the capitalist system improves, the government plays a more important role in the construction of urban infrastructure, the prevention of infectious diseases, the mediation of labor relations, and so on.

Fourth, the timely transformation of public policy and strong economic power is an important and secure way of dealing with social crises effectively. In the stage of rapid urbanization, a series of social contradictions are likely to emerge, and if they are not handled properly they will lead to the occurrence of social crises. The countries that developed first have experienced serious social and environmental problems, but strong financial resources and adjusting social policies and promoting social transformation in a timely fashion have helped them cope with such crises. The urbanization of countries that developed first reflected the characteristics of those times, and therefore do not necessarily provide a reference to those countries that are currently in the earlier stages of urbanization. However, taking the industrial revolution as an example of an opportunity to achieve a positive interaction between and development of industrialization and urbanization, yielding a reasonable return to government based on market mechanisms and adjusting public policies to promote social transformation still carry practical significance for those countries currently undergoing urbanization.

### 4.1.2 Fast-Pursuit Countries: Synchronized Industrialization and Urbanization and a Rapid Increase in the Level and Quality of Urbanization in a Relatively Short Period

After World War II, the world ushered in a new wave of industrialization and urbanization. South Korea, Israel, and other countries seized this opportunity, completing high-quality industrialization and urbanization in a relatively short period. The characteristics of industrialization and urbanization for this group of countries are as follows.

First, high-speed industrialization is a powerful driving force for the rapid development of urbanization. The periods from the 1950s to the 1970s for Israel and the 1960s to the 1980s for South Korea were characterized by a "squeeze type" of rapid industrialization, maintaining two-digit industrial growth rates for a sustained length of time, driving the average urbanization growth rate of 1.5 % points per annum and essentially completing urbanization in around 30 years. In this process, continuous industrial upgrading and a large amount of human capital investment promoted the competitiveness and labor productivity of South Korea and Israel. Industry continued to improve, forming a modern industrial system with global competitiveness, occupying the high-end of the global industrial chain and crossing the "middle income trap", successfully setting up a model for the developing countries.

Second, the government is an active promoter of industrialization and urbanization. Israel and South Korea are typical representatives of the "East Asia Model", in which the role of the government in the rapid industrialization and urbanization is very obvious. On the one hand, the government places great importance on land and space planning, playing an important role in guiding and optimizing the spatial distribution and layout of urban areas and promoting the rational allocation of resources. On the other hand, the government promotes industrial upgrading and strategic transformation through effective intervention. Japan has formulated policies for aspects such as tax, land, and infrastructure to support the development of high-tech industries. South Korea has actively implemented a government-led growth strategy and in recent years it has turned to the "national balanced development strategy", which provides a sustainable way for urbanization and economic growth to proceed.

Third, there is an emphasis on the overall planning of urban and rural development and agricultural modernization. South Korea has always attached importance to agriculture and rural modernization, having taken the improvement of agricultural productivity as the prerequisite and basis for promoting industrialization and urbanization. In the 1970s, South Korea implemented the "New Village Movement", greatly improving production and living conditions in rural areas, lifting the quality of life in such areas, raising the income level of farmers, and easing the enormous pressure caused by urban social transformation brought about by the rapid urbanization. The implementation also avoided rural recession through the improvement of rural infrastructure, culture, education, and other public services.

Fourth, effective measures have been taken to solve urban social problems and to achieve a stable social transformation. The rapid development of industrialization and urbanization has also brought Israel and South Korea some serious social problems. In the process of industrialization, South Korean enterprises adopted a management style that discriminated against workers, which caused national labor unrest to break out, prompting labor policies to be amended. In particular, the government became more neutral in the conflict between labor and capital. The government emphasized the legal and administrative means for regulating labor behavior and for establishing negotiation mechanisms and labor dispute mediation mechanisms so as to generate more harmonious industrial relations and working conditions.

Although Israel and South Korea still face their own problems with respect to urbanization, their achievements to simultaneously raise both the level and quality of urbanization, in a relatively short period, are remarkable. This achievement is based on their own natural resources and cultural traditions, effective government guidance and action, an emphasis on the integration of urban and rural areas and on agricultural modernization, and resolving in a timely fashion the complex social problems associated with urbanization.

### 4.1.3 Medium-Income Countries: Strategic and Institutional Defects Caused High-Level but Low-Quality Urbanization

Since WWII, there has been remarkable progress in the urbanization of most countries in Asia. However, the associated social and natural environmental problems are becoming more pronounced and are exerting a greater influence on the process of urbanization. In particular, a large number of agricultural workers have migrated into cities, and the urban poor in large cities have become a typical phenomenon of urbanization. The characteristics of this group of countries are as follows.

First, the respective governments have dealt with the economic and social problems of industrialization and urbanization ineffectively. Urbanization has lacked integrated planning, a large number of migrants have moved into cities in an unplanned and disorderly way, city centers have developed abnormally, and regional development has not been not balanced. In addition, attention to social issues has been insufficient, the gap between the rich and the poor has become huge, the problems of urban poverty and urban crime have become serious, and social conflicts have intensified.

Second, the historical land ownership and management system has not provided a solid foundation for urbanization. The traditional agricultural model and land system still occupy important positions. However, the agricultural sector remains relatively backward and lacks sufficient demand and power for industrialization, and economic development and the initiation of modern industry are almost entirely driven by the export of primary products. Furthermore, the majority of farmers without land find it difficult to obtain sufficient earnings from the export of primary products, and living conditions are harsh. Therefore, they can no longer base themselves in rural areas, and instead are moving into cities, resulting in a reduction in agricultural productivity and promoting depression of rural areas.

## 4.2   Lessons for China

Owing to strategic and institutional defects, middle-income countries have high levels of urbanization but it is of low quality. The experience of these countries and the lessons drawn therefrom allow us to appreciate the importance of the coordinated development of industrialization and urbanization, and of laying down a solid, substantial institutional foundation for the healthy development of urbanization (Wang et al. 2012). China falls in the range of middle-income countries, and also shows characteristics of the high level and low quality of urbanization due to strategic and institutional defects (Zhang 2016). China is now in the stage of rapid urbanization, and the nation should learn from the experience and lessons of the urbanization of the different Asian countries examined above, as follows.

First, urbanization should be adapted to the process of industrialization. We must pay attention to the supportive role given to urbanization by industrialization and to industrial development by urbanization. After WWII, South Korea achieved rapid urbanization by industrialization. The rapid industrial development provided strong support for increasing the number of cities and for expanding existing urban areas. Urbanization that occurs beyond the processes of economic development and industrialization is abnormal (Beaudry and Schiffauerova 2009). For example, because industrialization is seriously lagging urbanization in Mongolia, the cities there lack effective industry and infrastructural support, and an abnormal and excessive urbanization pattern has emerged. Problems associated with urban employment, housing, medical care, and education, amongst others, are difficult to solve, which is not conducive to social stability.

Second, there is a direct relationship between the development of a city's leading industry and the development of urbanization. The urbanization experience of Japan, South Korea, and Israel shows that developing the local dominant industry with competitive advantage is an important condition for the development of urbanization. When a city develops to a certain level, its ability to harness capital, labor, and other factors of production by itself is the decisive factor in urban development. The ability of the city to do this lies in forming a dominant industry with competitive advantage and can be sustainable development (Jong et al. 2015). Therefore, the key role of supporting industrialization through modern industry is to solve the problem of maintaining industrial development during rapid urban development.

Third, there needs to be integrated development of urban and rural areas (Hsieh 2014). Indonesia's urbanization is concentrated mainly on the island of Java, and development of other areas of Indonesia is slow, which has resulted in the formation of serious disparities and disequilibrium. Currently in China, urbanization development of the eastern coastal areas is relatively fast, but the development of the vast central and western regions is relatively slow, and the gap is getting wider. In eastern China, the population density is too high and natural resources are relatively poor; in the central and western regions, the population density is low and natural resources are rich. With the current distribution of population and resources, economic development is relatively slow. It is necessary to correct this uneven development, as well as to strengthen the central and western regions, to create favorable conditions for economic development in those regions, to continue preferential policies for those regions, and to attract a variety of funding and types of talent to those regions. By doing all of this, it is more likely that coordinated development of the whole country will be achieved. We also need to strengthen the cooperation between the eastern and western regions, including with respect to personnel, capital, technology, and industry.

Fourth, attention should be paid to the government's proper guidance (Henderson and Wang 2007). The problems exposed in the urbanization process in Indonesia reflect the lack of the government's macro-regulative and control functions. This problem is worth paying attention to, and the relevant government departments must give correct guidance regarding the process of urbanization. The experience of urbanization in Japan, South Korea, and Pakistan also highlights the importance of government guidance. Because urbanization involves economic structure adjustment, social structure change, and a suitable spatial arrangement, the government needs to offer macro-regulation and guidance. The process of urbanization is like an engineering system, requiring the coordination of all parties because the optimum of each subsystem does not necessarily represent the optimum of the entire system. The implication is that the government should take a whole-country perspective to promote the process of urbanization through effective planning and macro-adjustment, reflecting the government's will in the process of development. China has a vast territory and a huge population. The mode of urbanization should not be confined to a specific development model and we should strive to promote the development of urbanization based on innovative ideas and from the points of view of industry development, urban construction, institutional innovation, and policy adjustment.

Fifth, there needs to be an emphasis on optimizing the spatial layout (arrangement) of urbanization (Ramachandra et al. 2015). The main feature of the Japanese and South Korean urbanization patterns is the prioritized development of a small number of central cities. The unbalanced spatial distribution of the urban system has a long-lived and profound impact on the development of the regional economies within each country. For example, Japan's long-term problem is the excessive concentration in Tokyo and the spatial problem of development that is too dense and constricted. To solve these problems, both Japan and South Korea have encouraged the development of satellite cities, and fostered new urban growth points, which help to disperse both the population and the industrial agglomeration of metropolitan areas. This forms a city sequence of large, medium, and small with the combination of development poles and a hierarchical order, thereby promoting the coordinated development of urban and rural areas. China is currently in a period of rapid urbanization and industrialization. With regard to effectively coordinating the development of urban and rural areas and regional economic development, and avoiding the emergence of spatially dense and constricted development, the examples of Japan and South Korea can provide us with useful insights.

# References

Beaudry C, Schiffauerova A (2009) Who's right, Marshall or Jacobs? The localization versus urbanization debate. Res Policy 38:318–337

Henderson JV, Wang HG (2007) Urbanization and city growth: the role of institutions. Reg Sci Urb Econ 37:283–313

Hsieh S-C (2014) Analyzing urbanization data using rural-urban interaction model and logistic growth model. Comput Environ Urban Syst 45:89–100

Jong Md, Joss S, Schraven D, Zhan C, Weijnen M (2015) Sustainable/smart/resilient/low carbon/eco-knowledge cities; making sense of a multitude of concepts promoting sustainable urbanization. J Clean Prod 109:25–38

Ramachandra TV, Bharath AH, Sowmyashree MV (2015) Monitoring urbanization and its implications in a mega city from space: spatiotemporal patterns and its indicators. J Environ Manage 148:67–81

Wang H, He Q, Liu X, Zhuang Y, Hong S (2012) Global urbanization research from 1991 to 2009: a systematic research review. Landsc Urb Plann 104:299–309

Zhang XQ (2016) The trends, promises and challenges of urbanisation in the world. Habitat Intl 54:241–252